The Collapsing Bubble

Growth and Fossil Energy

Lindsey Grant

SEVEN LOCKS PRESS

Santa Ana, California

Seven Locks Press
P.O. Box 25689
Santa Ana, CA 92799
(800) 354-5348

Second printing.

Individual Sales. This book is available through most bookstores or can be ordered directly from Seven Locks Press at the address above.

Quantity Sales. Special discounts are available on quantity purchases by corporations, associations, and others. For details, contact the "Special Sales Department" at the publisher's address above.

Printed in the United States of America

Library of Congress Cataloging-in-Publication Data
is available from the publisher
ISBN 1-931643-58-X

CONTENTS

FIGURES AND TABLES

Introduction

For three years, the nation has been obsessed with the threat of terrorism, but the energy transition, which we are now entering, poses a far vaster threat to our well-being and lives than wars or terrorists. For two or three generations, reliance on coal will generate much more serious environmental and climate problems than petroleum or natural gas. How will the world "sequester"more than 16 billion tons of carbon dioxide each year until the coal runs out? After that, we will enter an age of "overshoot" because the populations that can be supported with fossil fuels are much larger than those that renewable resources can support.

I have assembled here three NPG FORUM articles I wrote this year for Negative Population Growth. The three fit together naturally. Together, they constitute (1) a critique of the faith in unending growth that animates our political leadership, (2) an examination of just how much fossil fuel may be left and of the hopes and dangers that attend its use, and (3) a survey of the renewable sources of energy that will be available as the fossil fuel era winds down.

The conclusion I reach seems to me moderately hopeful: that renewables can provide for future generations' energy needs, but only if U.S. population has begun a sharp descent in this century—rather than continuing its present wanton growth—and only if we begin very soon to manage the transition. The real cost of energy will be much higher than it has been in the brief era of fossil fuels, and fundamental changes will be needed as we lose the convenience of petroleum. Even the sanest of approaches must negotiate some frightening environmental and climate threats. But if we continue with "business as usual," we will blunder into calamities.

That, I may add, is not likely to sound optimistic to growth enthusiasts. Their theme is the Mardi Gras theme: "Let the good times roll!" But the calamities will multiply if they have their way.

My focus is on the United States, though I discuss global statistics at some length because we have become interdependent. If I assumed "energy independence," this report would be much bleaker. Other nations will need to make their own calculations, and most of them face problems much more serious than ours.

These three papers focus on specific and concrete evidence of where we are and where we should be heading. For my view of the broader context, I refer readers to *JUGGERNAUT: GROWTH ON A FINITE PLANET* (Seven Locks Press: 1996). For a much more detailed philosophical treatment of the problem of excessive growth, I would go back perhaps to William Catton's *OVERSHOOT* (Illinois University Press: 1980).

The energy transition is the most proximate threat to the industrial world. For much of the rest of the world, food and water are the immediate and overwhelming problems. The issues are tied together because modern agriculture depends on massive inputs of hydrocarbons. I will try to quantify that connection in Chapter 3.

— Lindsey Grant
Santa Fe, NM, October 2004

Chapter 1.
The New American Century?

"Anyone who believes exponential growth can go on forever in
a finite world is either a madman or an economist."

—Economist Kenneth Boulding

***

In 1997, a small group of neo-conservatives organized the Project for the
New American Century (PNAC) and published a Statement of Principles.
The organization's name describes its state of mind. Its philosophy is per-
haps most succinctly expressed in a short quotation from the Principles:

> "We need to accept responsibility for America's unique role in
> preserving and extending an international order friendly to our
> security, our prosperity, and our principles. Such a Reaganite
> policy of military strength and moral clarity may not be fashion-
> able today. But it is necessary if the United States is to build on
> the successes of this past century and to ensure our security
> and our greatness in the next."[1]

Among the signers of that declaration were our Vice President Cheney,
Secretary of Defense Rumsfeld, and Deputy Secretary of Defense Paul
Wolfowitz. The group is proud to claim parentage of the administration's
Iraq strategy. Its views are reflected in the White House position that there
is "a single sustainable model for national success: freedom, democracy,
and free enterprise... These values of freedom are right and true for every
person, in every society . . . and the duty of protecting these values
against their enemies is the common calling of freedom-loving persons
across the globe and across the ages."[2] Condoleezza Rice called this our
"moral mission," and it was cited as an argument for invading Iraq.

The Project believes that "a cheap energy policy will lead to sustained, rapid, long-term economic and employment growth."[3] The Bush administration agrees. In April 2003, Under Secretary of State Alan Larsen told the Senate Foreign Relations Committee that the United States must have access to energy "on terms and conditions that support (our) economic growth and prosperity," and that we require "improved investment opportunities" in the energy producing regions of the world.[4] It would, as the saying goes, be nice if you can do it, but Mr. Larson has his work cut out for him, as I will demonstrate later.

The neo-conservatives have belatedly learned that petroleum is indeed finite, and that we are running out of it. The Iraq invasion bears all the marks of a deliberate, and failed, policy to take political control of a country in the middle of the oil patch so as to assure our future oil supplies. If free trade and investment are a mask for taking control of the resources we want, it will encounter mounting resistance from others, such as the recent riots in Bolivia that forced the government to back out of a contract to sell gas to the United States.

Even Canada may begin to wonder whether, in a world increasingly desperate for fresh water, the U.S. may eventually demand a share of Canadian water resources—despite the Canadian policy prohibiting the bulk export of water.

The assertion that we must assure our access to others' resources runs in uneasy harness with the belief in growth. President Bush (like other Presidents before him) has called for faster economic growth. In a bid for the Hispanic vote (perhaps a misdirected bid), he has proposed a program to legalize illegal immigrants and to allow businesses to import more labor when they claim to need it, both of which will dramatically accelerate U.S. population growth—and our appetite for resources. And the Democrats are trying to outbid him.

Taken together, the assertion of our moral rectitude, our right to impose our values on the world, the desirability of continuing growth, and our right to support that growth with access to others' resources constitutes a sweeping assertion of our rights and power that seems a bit ambitious for a country with annual budgetary and foreign trade deficits of about $500 billion. Foreigners are financing that trade deficit—more than $1 billion a day—and buying out our businesses, while net U.S. capital investments abroad decline dramatically. But there are bigger problems than that.

In this paper, I will argue that the coming century is more likely to be a debacle than an American hegemony unless we curb our spendthrift ways, stop and reverse U.S. population growth, and help others to control theirs.

GROWTHMANIA: THE DURABLE BUBBLE

Before the 1300s, the idea of unlimited growth hardly figured in human thinking. Then came the Renaissance, which led to the Age of Exploration to the new world and new wealth. It started the agricultural and industrial revolutions and set in motion a worldwide scientific enterprise that is still accelerating. It began a period of enrichment and growth without parallel in human history.

The period has lasted, with minor interruptions, for six centuries. That success led gradually to a widespread conviction that growth is the natural and desirable order of things, and forever benign. Enter the Romantic Era and its sense of limitless horizons, and the "Age of Exuberance" (to borrow William Catton's term). Western civilization is still in that mode and is teaching it to the East and South.

It is a formidable belief system, but its proponents have forgotten that Its origins were not in population growth, but in the Black Death, the most widespread and severe population collapse in human history. Brutally, it readjusted the ratio of people to land. The surviving peasants found themselves with more farmland and more wealth. New wealth flowed into the depopulated cities. The institutional constraints of Feudalism were swept away and replaced by the system now identified with capitalism.[5] The subsequent Age of Exploration further improved the ratio of land to people by opening access to the new world, which has more than quadrupled the arable land available to Europeans.[6]

THE GROWTH MACHINE

One legal innovation, the limited liability corporation, was fundamental in promoting and shaping the age of growth. It changed the calculus of risk. If you succeed, unimaginable wealth. If you fail, you lose only the money you had put in the company. It was an immense inducement to risk-taking, an astonishing engine of growth, and the vehicle for the rise of capitalism.

Capitalism is uniquely the system for the entrepreneur, the risk-taker, the business adventurer. It serves the successful. So do its theoreticians. Conventional current economics is grounded in the expectation of endless growth. Economic growth for more profits. Population growth for more markets and cheap labor. (Not economic growth per capita, which would be more reasonable.) The economists' other myths and simplifications—economic man, infinite substitutability, comparative advantage, free trade and investment—all justify the freedom of action of the corporation.

The "economic man" hypothesis assumes that people displaced by change will find other and probably better employment. The overwhelming current evidence is to the contrary. "Infinite substitution" is regularly argued but never proven. It justifies the faith that growth can go on forever. (Right now, with an accelerating fresh water crisis, one may reasonably ask: What is your proposed substitute for water?)

Free trade is said to maximize efficiency through comparative advantage. It also widens the playing field for the TNC (trans national corporation), as does the prospect of unfettered investment and movement of capital.

The Austrian-American economist Joseph Schumpeter recognized that there is much suffering as old arrangements are swept aside, but he characterized it ingeniously as "creative destruction"—the old must be swept away to make room for the new and efficient. It is devil take the hindmost, unless it is controlled by social restraints, which themselves may be blocked by the financial and political power of capital.

Politicians respond to the siren song, and so do investors. Even now, as the world and the country try to shake off a recession, investors do not ask, "Were we on the wrong course?" but rather, "Is the slump over? Can we start coining money again?"

THE MIGHTY ENGINE WITH NO BRAKES

In the twentieth century, modern medicine and public health programs lowered mortality in the poorer countries, and modern agriculture fed the rising numbers, but too little was done, too late, to lower fertility. That created a fundamental demographic imbalance. The resulting population growth has dwarfed all previous human experience. World population quadrupled in one century, a change so astonishing that it has altered—or should have altered—our assumptions as to the human connection with the rest of the planet.

Are we plunging toward a collapse because of that very success? Philosophers since John Stewart Mill have warned against the illusion of perpetual growth. Endlessly growing numbers cannot enjoy endlessly growing consumption. There is a mathematical platitude that post-Keynesian economists ignore: Material growth at some point becomes a logical impossibility on a finite planet. When?

John Maynard Keynes is something of a demi-god to conventional modern economists. When the machine stopped in the Great Depression, Keynes offered a way to start it again. However, Keynes was not as wedded to growthmania as his followers. He raised serious questions: Can growth go on indefinitely? Would it be desirable?[7] Is market capitalism—motivated by greed—a sound moral basis for society?[8] Those doubts were swept aside in the rush to profit.

Herman Daly, considered a renegade by conventional macroeconomists, makes a point his colleagues ignore: The economy is a subset of the environment; it is not independent. The Earth is not simply a source of resources and a sink for the waste products—the principal products—of economic activity. It is the matrix that sustains life, including human life, and we must ask whether human economic activity is degrading that matrix.

Two hundred years ago, Thomas Malthus worried (perhaps prematurely) about how many people the Earth could support, but he did not ask the next question: What will increasing human numbers do to the Earth? In 1864, George Perkins Marsh was the first to systematically address that question.[9] Science has been describing the impacts ever since. In 1992, the Presidents of the U.S. National Academy of Sciences and the British Royal Society adopted a joint statement (later adopted by the world's major national scientific societies) that, "If current predictions of population growth prove accurate and patterns of human activity on the planet remain unchanged, science and technology may not be able to prevent either irreversible degradation of the environment or continued poverty for much of the world."[10] If expanding populations and growing consumption impose unbearable strains on the ecosystems that support us, we must learn to identify the turning point and ask: What population is sustainable?

Population growth is not necessary for well-being. Japan and Europe, with stable or declining populations, show a vitality that belies the common wisdom. A Brookings Institution study examined cities' growth and prosperity in depth. It concluded that, "We have punctured one important piece of conventional wisdom: the idea that achieving income growth in a metropolitan area requires population growth."[11] Various other studies have shown that the residents in stable cities are likely to be better off than those in rapidly growing ones, both by economic measures and quality of life indicators. Pittsburgh, Pennsylvania, long the epitome of the "rust belt," ranks at or near the top on both scales, despite two generations of population decline—and despite its wretched weather. And taxes tend to rise with urban growth.

The literature challenging growthmania has itself been growing, documenting the charge that the benefits of growth have gone to the entrepreneurs rather than to the mass of working people, and that the growth of the human economic enterprise has run down the natural capital of the Earth—which does not appear in GNP statistics.[12]

Mainstream economics has ignored that literature. In the pursuit of growth, it has brushed aside every doubt.

The enthusiasm for population growth is hardly universal. President Nixon asked whether it was a good thing. He persuaded Congress to create the Commission on Population Growth and the American Future (the "Rockefeller Commission"), which concluded that it could see no advantage in further growth of the American population. Unfortunately, President Nixon shelved it for political reasons (and so have all subsequent Presidents). That was thirty-two years ago. We have added 86 million people since then.

Polls suggest that the American public is not enamored of further population growth, but there is a virtual political taboo. Almost nobody mentions the demographic consequences when politicians discuss policies such as increasing immigration that generate population growth, because the pro-growth argument is endorsed by the powerful.

Nevertheless, growth must stop. The question is, when and where will it stop?

THE MEASUREMENT OF OPTIMUM POPULATION

NPG examined the concept of optimum population fifteen years ago, in a series of FORUM papers. Populations, U.S. and worldwide, have grown substantially since then, as has the addiction to growth among our political leaders. Perhaps it is time to revisit the concept in light of the developments in recent years.

The effort to define "optimum population" challenges the prevailing economic and political wisdom that growth is, by definition, a good thing. So be it. The challenge itself is at least as important as the number that we may eventually assign to optimum population.

Maximum population is simply an estimate of how many people can be supported at a given time. Sustainable population is the population that can be supported, indefinitely, without degrading the ability of the ecosystem to support it. Optimum population extends that idea; it undertakes to describe a population level that could live a comfortable life within those resource and environmental constraints. It is the antithesis of current economic goals, but it should be congenial with the economic aspirations of all but the greedy. And it is a vision of a future without the threat of collapse.

Putting numbers on optimum population is a mix of science, value judgments, and outright guessing. How do we decide whether further population growth is bad and what numbers would serve humanity better? I will briefly identify a few such yardsticks below.[13]

Food. World food production kept ahead of population growth in the 1960s and 1970s, stayed just ahead in the 1980s, and fell behind in the 1990s. Grain production has been stagnant since the mid-1990s, and even that may not be sustainable. Our hope for higher yields now rests mostly on genetic modification (GM), itself a dangerous project.

The other sources of rising yields are beginning to fail. Chemical fertilizers produce less and less additional food as yields rise. Eventually, the added fertilizer does not pay for itself. The developed world has passed that point, and China is approaching it.

Modern agriculture depends on petroleum and gas to run its heavy machines and provide feedstock for fertilizer plants, and we are approaching an era when both fuels will be in short supply.

Arable acreage is declining and topsoils are eroding. As a result of population growth and urban sprawl, arable land per capita has declined since

1970 by one-third (to 0.16 hectares) in the less developed countries, by one-fifth (to 0.2 hectares) in Europe, and by one-third (to 0.63 hectares) in the United States. Thirty years.

That acreage figure for the United States points to a subsidiary lesson. We still have more room than most countries. But our rapid growth narrows the advantage. We supply one-third of the grain that enters international trade, but if yields and our consumption habits stay as they are, we will need that grain ourselves in one generation (assuming the Census high projection) or two (assuming the middle projection). It will take a remarkable increase in grain yields, plus a dramatic dietary shift away from meat, to feed our own growing population through this century, to say nothing of exporting grain. And such increases in yields seem most unlikely in the face of the constraints I described earlier.

The chemical industry will compete for more and more land as it turns to cellulose to replace hydrocarbons as feedstock, and as crops are engineered through GM to produce pharmaceuticals and other chemicals.

Irrigated land now produces 40 percent of the world's crops, but salinization is lowering yields in perhaps one-third of world-irrigated cropland. Irrigation uses about 70 percent of human fresh water consumption, but we are running out of water. Rivers are going dry and water tables are declining in China, India, Pakistan, the Middle East, Mexico, and the American West. Even moister regions are feeling the pinch. Freshwater data are notoriously inexact, but a United Nations study in 2003 found that global per capita water supplies declined by one-third between 1970 and 1990 and are likely to decline by another one-third in the next twenty years, and very little is being done about it.

Climate change threatens food production (see below).

The world has run through the windfalls provided successively by the Black Death and the opening of the new world. Much smaller populations, with more land per capita, would provide a cushion against the threats to food production.

Modern agriculture is itself destructive. The world now uses about six times as much commercial fertilizer as it did in 1950, and twenty-five times as much chemical pesticide. Human activities put nitrogen compounds, potassium, phosphates, and sulfates into the environment faster than natural processes produce them, and we are just beginning to understand the consequences. Monocultures and high-yielding "green revolution" crops demand more water and more pesticides. New pesticides are introduced as pests develop resistance to the old ones. It is a squirrel cage, and experts differ as to whether it has reduced the proportion of the crops that are lost to pests.

World food production could be sustained at roughly half its present level with a judicious combination of organic manures and chemical fertilizers. (Just before the reliance on commercial fertilizer, U.S. corn yields were about 40 percent of current yields.) We would need to change our ways and utilize more natural manures from livestock and, indeed, from humans, but that in itself would solve some serious pollution problems. Very roughly, half the production would support half the present population, and it would be much less damaging agriculture.

Health. The proliferation of chemicals is not just an agricultural problem. There are four times as many chemicals in the world chemicals registry as there were in 1980. We all carry hundreds of those new chemicals in our bodies. Some of them are known sources of cancer, endocrine disruption, immune system suppression, falling sperm counts and infertility, and learning disabilities in children. And most of them have not been tested for their impact on health or the environment.

The urban population in the less developed world (LDCs) was 300 million in 1950. By 2000, it had reached two billion, propelled largely by desperate peasants moving to cities to stay alive. Water supplies, sewage services, and electric supplies have lagged far behind, and it is remarkable that the crowded slums have not generated more epidemics than they have. With the public health measures that kicked off the population explosion now in disarray, rising mortality may forestall the United Nations' (UN) projection that LDCs' urban population will reach four billion by 2030.

The growth of cities and growing water shortages mean that city residents in the less developed countries reuse sewage, with disastrous health effects. Even in the industrial world, sewage plants filter out the solid wastes and kill the microbes but usually leave the nitrogen in the water; and we have not started to try to filter out the many drugs that people take and then pass on to others. They can be detected even in rivers below the sewage outfalls.

It would be a happier world with fewer chemicals and better water.

The Microbial World. This chemical assault affects other animals. It may be endangering the microbes that we depend upon but cannot see. For one example: Earth microbes have so far converted nitrogen fertilizers back into inert molecular nitrogen fast enough to keep us from swamping the Earth in nitrogen compounds, yet we don't know how much of a chemical load the microbes can tolerate.

Human population growth drives chemical production, both by keeping up the pressure for more food production and by increasing the demand for non-agricultural chemicals. A much smaller population would lead to a reduction in the introduction of chemicals into an environment which we are changing but do not really understand.

Fisheries. Worldwide marine fish production rose from 20 to over 70 million tons from 1950 to the late 1980s, but has stuck there. Then came a soaring growth in aquaculture, which pollutes the water, competes with livestock for feed, and concentrates the harmful chemicals we are putting into the environment. (The Environmental Protection Agency recommends eating one serving or less of farmed salmon per month.)

It would be a better world if human demand for fish and the pollution we put into the ocean were both closer to the 1950 level.

The Energy Transition. Fossil energy is a profound disturbance to the ecosystem. It moves carbon—and sulphur, arsenic, mercury, chromium, lead, selenium, and boron—from the lithosphere into the biosphere and the atmosphere, at a rate and scale greater than all natural processes. We worry about the threat of terrorism to petroleum supplies, but the supply will decline, anyway. That will be an environmental boon but an economic disaster, unless we have prepared for it.

Estimates of the world's remaining petroleum resources range around two trillion barrels.[14] World consumption is presently about 28 billion barrels a year. Dividing the estimated resources by current annual consumption, it is commonly (and erroneously) said that about seventy to eighty years' supply remains, but consumption is rising fast, as China and India industrialize.[15] Not a very long future.

United States crude oil and natural gas production peaked over thirty years ago. The country now produces 40 percent less crude oil and 13 percent less gas than it did then. U.S. petroleum imports account for 62 percent of our consumption now, and the proportion is rising.[16] With less than 5 percent of the world's population, we consume 26 percent of world petroleum production. The share is going down as others—including the rising giants, China and India—compete for a larger share.[17] China is moving into a stronger bidding position than ours because it is not saddled with massive trade deficits. Under Secretary Larsen's vision of the United States moving in to exploit others' petroleum resources may be an anachronism.

Those who expect continuing growth in petroleum consumption ignore petroleum geologists' warnings that world production will begin to decline, probably in less than twenty years. Extracting the remaining petroleum will become more costly, competition for petroleum will intensify, and prices will rise sharply. Gas will follow petroleum. Not a happy prospect for a nation that is already by far the biggest importer and wants to import more.

Repeated military interventions to secure oil will become less and less effective because of mounting resistance abroad and rising discontent in this country over the financial and moral burdens and the military appropriation of civilian oil supplies. A vast and sophisticated military that has to fight abroad for the oil it needs to operate is a costly and uncertain tool.

American politicians have regularly talked of "energy independence" even as we have grown more and more dependent on foreign sources. (Who wants to be dependent on the unstable Persian Gulf?) We won't get back to the good old days in petroleum, even if we get population growth under control, but it would help our adjustment to a new and leaner energy mix.

Coal is more abundant, and much of it is in the United States, but it is a dirty fuel. Some of the pollution could be controlled at a high cost, but the carbon dioxide, and its effect on climate, is a particular problem.

Growth apologists look for panaceas. They suggest oil sands and shales, but processing them is environmentally destructive and may demand more energy than they would yield. Ocean methane from the continental slopes is suggested, but the environmental consequences could be frightening.[18] The activity might release the methane without capturing it, thus further warming the climate and triggering undersea mudslides and tsunamis.

Biomass is a very limited solution because its production competes for land with rising human needs for food and cellulose.

Wind and photovoltaics can only supply electricity, while petroleum has been used for everything from airplane fuel to chemical feedstock. For peaking power, wind energy is nearly competitive right now, and much more benign than fossil fuels. For reliable base power, however, wind and solar energy will be much more expensive than fossil fuels are now because of the problem of storing the energy until it is needed.

The world is headed into an energy transition, probably toward a mix of coal, nuclear, and more benign renewable power. The rising costs and dislocations will threaten the world's economies. A saner U.S. policy would stop the effort to monopolize other countries' oil supplies and instead look toward reducing our demand for petroleum and gas. We must phase out our current waste and, more fundamentally, we must stop and reverse current population growth. A smaller population would make the energy transition easier, but demographics move slowly.

Climate Change. Fossil fuels generate climate change, which is beginning to reduce crop yields, especially in the poorer countries. It is already raising sea levels and generating more extreme weather: floods, droughts, extreme hot or cold spells. The impacts are likely to worsen for centuries. So far, the human race is doing very little about the problem it has created.

In 1995, the Intergovernmental Panel on Climate Change (IPCC) estimated that it would take an immediate reduction in carbon emissions to 30–50 percent of present levels to hold the human impact on climate even at its present levels. In the face of that calculation, the modest reductions proposed in the Kyoto protocols are largely symbolic.

Population size must be addressed if we are to come close to the 30–50 percent goal. With populations at 1950 levels, the world would have been within that range even with present per capita emissions.[19] Whatever we can gain in energy efficiency would be lagniappe.

Technology, the Headstrong Servant. Modern Americans expect technology to solve our environmental problems, but it actually generated most of them. It can be of help. When the United States Government passed the Clean Air Act in 1972, technical fixes reduced some of the principal pollutants. Technology has its limits, however, and overall air pollution has been rising again for several years. Technology can be part of the solution, but not all of it. Lower numbers and lower demand are central to reducing pollution.

In one respect, technology has betrayed the pro-growth economists. They call for economic growth to provide jobs for growing populations. But technology has driven productivity up. Economic growth is not necessarily job-connected anymore, as we have been learning in recent years. Businesses can turn to automation, instead. The solution for unemployment and low wages is fewer workers competing for jobs. Proponents of more immigration, take note.

Social Equity and Human Numbers. China and India explicitly seek to raise per capita income to the present average level in the industrial world, and most poor nations would probably agree. The effort to get rich has created horrendous pollution problems in China. If the poor countries are to get as rich as they hope, without increasing gross world economic activity and further damaging the world's environment, world population would have to be not much over one billion.

Non-linearities. My analysis so far has been linear (i.e., so much more of a given input produces a comparable change in the impact). In fact, nature is seldom linear. In the study of climate change, for example, scientists are regularly identifying non-linearities—feedback loops that may intensify the prospective problems—from alterations in ocean currents which could alter weather worldwide and make Europe's climate like Labrador's, or the warming effect of diminishing ice and snow fields, to the release of stored methane from the ocean and carbon dioxide from Arctic tundra.

Prudence would suggest that we not press our present systems to the limit, so that we may have space to maneuver if unexpected changes reduce the productive capacity of our support systems.

Crowding and the Intangibles. I was told of a kid from the New York City ghetto who was sent to a city-owned summer camp in the hills. The bus arrived after dark, and when the kid stepped out, he looked up and said, "What are all them little white things up there?" He had never seen the stars. That, I submit, is deprivation. It is getting worse as cities grow and the sky gets murkier.

The kid was hardly unique. There are literally thousands of newspaper stories about the strains of increasing crowding in the United States.[20] In some degree, they result from our insistence on a costly and inefficient lifestyle, but they are even more fundamentally the product of population growth. We don't like to be crowded, nor do the people of more crowded lands who have become resigned to it. The search for optimum population should include the calculation: How much room do we like?

THE BOTTOM LINE

The United States' power and well-being rest on flimsier footings than the Administration and the New American Century members seem to believe. Our balance of payments deficit is chronic and worsening. If foreigners turn away from the United States as the residual safe repository for their funds, it will drive the dollar down, fast and far. That might encourage what exports we have left, but it would generate massive cost-push inflation. Our budgetary deficit contributes to such a scenario. The current decline of the dollar may be a harbinger.

Those problems would be manageable, if the United States had the discipline and the will. The issues of food, water, energy, health, climate, and crowding are more fundamental and can be addressed only if we abandon our fixation on growth and address the demand side rather than denying it is a problem. Each of those issues can be resolved only if we move toward a smaller population.

My fellow writers on optimum population would probably agree that for the United States, optimum may be something like the numbers we passed around 1950: 150 million (half the present 293 million), give or take a half.

It is much harder to put a number on optimum world population, because an outsider can hardly determine what consumption level might seem "comfortable" for the billions of people who are presently at or close to the margin of survival, and we cannot know what tradeoffs countries will choose between prosperity and pollution. Perhaps the 1950 figure of about 2.5 billion (40 percent of the present 6.4 billion) would be an upper limit. I have pointed out that, for everybody to achieve something like the average consumption level of the industrial world without vastly increasing pollution, world population would have to be in the vicinity of one billion.

The poor countries arrived too late to join the feast. Most of them have little or no fossil fuel and no hope of enjoying a boom period based on the rapid draw-down of a one-time energy source, such as the industrial world enjoyed. Their arable land is over-used and deteriorating. They suffer the most from water shortages and climate change. The problems of poverty and the competition for resources are producing tensions and conflict, whether they take the form of intensified migration to the West, or terrorism, or rogue states or interminable local wars and insurrections. For them, a future with far fewer people and more resources per capita would be a much happier future. Again, I think of the unexpected results of the Black Plague, though I would hope for a more benign process.

The less developed world has grown by two-thirds since 1950—and they were poor in 1950. The need for a fundamental shift in the ratio of resources to people in the poor countries may itself justify an optimum world population figure of one billion. Barring a catastrophe, it might take centuries to reach such figures, even with a determined worldwide effort.

Europe and Japan are already on the way to lower populations and must face the question: Where should they stop? I'll come back to that.

WHY SUCH ROUND NUMBERS?

Those are hardly rigorous calculations. There are too many horseback calculations and value judgments. What living standard is "comfortable"? There will be unpredictable technological changes, and continuing environmental degradation will almost certainly diminish the Earth's support capability.

But then again, when do we know the exact consequences of any major decision? They are all made on the basis of partial information, and they can be refined only as we go along and learn more. Precision is not required here. If the weight of evidence suggests that a population should be smaller than it is now, the policy implications are similar, whether the gap is 100 million or 200 million. The important thing is to ask the question, in one context after another, would this problem be more easily solved with a smaller population or a larger one? I think the examples above provide the answer.

WHY TRY TO ESTIMATE OPTIMUM POPULATION?

We need to show that human numbers matter in order to illuminate the flaws in growthmania. When I point out that a given policy will lead to more population growth, a typical reaction is "so what?" The present debates about immigration, welfare, and tax policies ignore their demographic impact and thus dismiss the future.

Defining a desirable population level is one step toward a more stable and less uncertain future. It sets the stage for the next necessary step: putting policies in place that will move human numbers in the right direction.

RESTORING A FLICKERING VISION

A vision flickered briefly in the 1960s and 1970s: It should be possible to combine modern technology with population stability, and thereby create a world in which all can live well. Modern productivity would replace the arduous physical toil whereby the poor labored to support the rich. That vision is being eroded because, in much of the world, economic growth is being absorbed by population growth that eventually eats up the gains.

Salvation may come from an unexpected source: young women with jobs and their own income and control over their decisions about childbearing. They have learned to practice family planning. In theory, that offers a way to regulate the balance between people and resources humanely, rather than through the grim operation of mortality as happened in the Black Death. In fact, women's choices have been based mostly on personal considerations, not on social or demographic grounds.

So far, in Europe, Japan, South Korea, Taiwan, and Singapore, and among non-Hispanic Whites in the United States, women have chosen to have far fewer children than would be necessary to replace themselves. In none of them has the fertility decline yet brought population down to optimum levels, but there are dramatic population declines in prospect if fertility does not soon rise to replacement levels. In many less developed countries, fertility is also declining, but not so far. It is not happening in Africa or the Middle East.

The world is tending to divide into two different demographic regions. In one of them, there is a real option of consciously managing population levels, but a need to define optimum population as a social goal and to enlist young women's participation in pursuing that goal. In the other, population growth is on a path that will stop and turn around only through catastrophe, hunger, and rising mortality.

HOW FEW IS TOO FEW?

For those countries poised at the edge of population decline, the question arises: How far? Who will support the aged? Is free trade a serious possibility when wealthy and aging countries' labor faces the competition from overpopulated poor countries, working for a fraction as much money? What does women's independence bode for the traditional conjugal family? (More than half of Swedish children are now born out of wedlock, and other European countries are not far behind.)

There are answers to those questions, but the more fundamental question is one of numbers. Italy's population will be eight million and still declining in 2100 if present fertility levels persist. If fertility should come back to replacement level by 2020—70 percent above present fertility—population would stabilize at 25 million, about 43 percent of the present level. Environmentally, it might be a good level. Practically, there are problems.

Will Italian women fit their childbearing to social needs? If not, how much immigration can Italy sustain? Genetically, Italians would progressively disappear, to be supplanted by the descendants of the immigrants. In the absence of action on fertility and migration, Italy could simply be overwhelmed by illegal migration from desperate countries to the south.

Population writers have yet to address the question, what is a desirable lower limit to optimum? Certainly, one cardinal rule is that fertility must at some point come back up to replacement level.

THE IMMEDIATE TASK

The more pressing task is to define and popularize the idea of an upper limit, and to act on it. For the United States, that would mean limiting immigration and persuading mothers to stop at two children, at least until growth turns around. We should return to the policies—largely abandoned during the Reagan administration and this one—of helping the poor countries to stop growth, which most of them want to do. They would be better off, and so would we, if they were not made desperate by growing idle and hungry populations.

Most poor countries know they are already too big, though none have adopted a target for optimum population. The United States is unique. Facing undiminished population growth driven mostly by immigration, we do not recognize the problem. We need to help the poor countries to accomplish their demographic revolution—and to apply the lesson to the United States.[21]

Chapter 2.
How Long the Twilight?

This country and the world are in for profound change as the petroleum boom winds down. I find that even specialists in the fields that will be most affected have not seriously considered what that transition will be like or how they will handle it. This book is an effort to describe the transition and to explore what lies beyond it.

In this chapter, I will examine the period of decline of petroleum and gas, which will be swift. The petroleum era has been a brief spike that has contributed to a quadrupling of world and U.S. population and rising consumption levels. We are entering an age of overshoot. There may be opportunities for an orderly withdrawal, if we are wise enough to manage the environmental threats and unlearn the faith in growth that has developed in the fossil fuel era. There will be disasters if we do not. Chapter 3 will look at a much more speculative future beyond fossil fuels and suggest that current populations cannot be supported without them. We may come to see the Industrial Age as the most intense human disturbance of our natural support systems in history. With the judicious employment of the technologies we have learned—and with a bit of luck—we may be able to create a more harmonious balance with the rest of the biosphere, but at much lower population levels and less consumptive habits.

***

THE "BUSINESS AS USUAL" SCENARIO

In the old journalistic tradition, I will summarize the projections at the beginning and explain them later. Figure 1 is a stacked graph showing the history of U.S. conventional energy consumption and a speculative projection of its likely path in this century, based on current trends and assuming no fundamental policy changes (such as those I advocate). Superimposed on it is a line showing past U.S. population growth and the Census Bureau middle projection of future growth.[1]

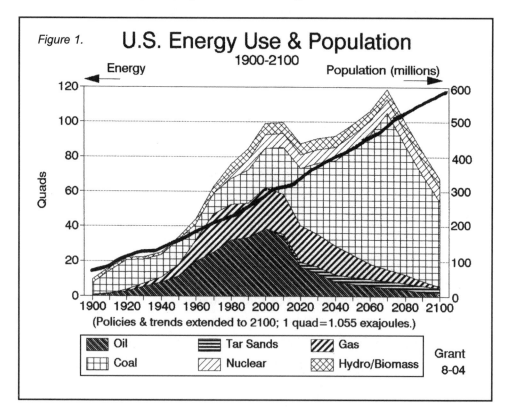

Figure 1. **U.S. Energy Use & Population**
1900-2100

(Policies & trends extended to 2100; 1 quad=1.055 exajoules.)

| Oil | Tar Sands | Gas |
| Coal | Nuclear | Hydro/Biomass |

Grant
8-04

The decline of petroleum and gas is pretty much out of our hands. Fission energy is limited by uranium resources, and hydropower by a lack of sites. With a growing population, biomass energy has little room for growth. The key is coal. There is enough coal to support a 2 percent annual growth rate for most of the century, but it raises major environmental concerns not shown on the graph. Nor does the graph deal with the evidence, which I will present in Chapter 3, that a sustainable future beyond the transition will require a population much smaller than now. The graph is optimistic in several respects. It assumes continuing petroleum imports and rising gas imports, and there are no surprises or interruptions of international energy trade.

THE PETROLEUM SPIKE

World crude oil production rose from negligible in 1900 to about 80 million barrels per day (mbpd) a century later. U.S. consumption of crude oil and natural gas have both risen more than 100-fold since 1900.[2] In that period, world and U.S. populations nearly quadrupled, but the dramatic per-capita increase in petroleum use lies at the heart of "The American Century."

The age of reliance on fossil fuels has been extraordinary, both for its swift rise and its prospective brevity. It has supported a remarkable growth in prosperity in the industrial world. The return to reliance on the sun's annual radiation of energy to the Earth will be a painful comedown.

Our political and business "leaders" seem generally oblivious to the unique character of the fossil fuel age. They consider growth the natural and desirable order of affairs and call for more of it—an outlook influenced more by greed than reflection. When warned of the brevity of the fossil era and the dangers it is creating, they defend the status quo or, when pressed, offer simplistic panaceas such as the hope that hydrogen or wind and solar energy will solve our problems. By themselves, they will not.

Fossil and Renewable Fuels in the U.S. In the United States, we consume 97 quads (quadrillion British Thermal Units) or 102 exajoules (quintillion joules) of commercial energy each year. Of that, petroleum furnishes 39 percent, natural gas 24 percent, and coal 23 percent. Taken together, fossil fuels contribute over 86 percent of our total energy. Nuclear power, which is a fossil fuel in the sense that it relies on ores from earlier geological times, provides another 8 percent. Renewable energy provides just 6 percent, almost entirely from hydroelectric power and biomass; wind energy provides 0.1 percent, and solar electricity—which is presently much in vogue—much less than that, or 1/1500th of the total.[3]

Fossil Fuels' Role in the Economy The U.S. economy is built on fossil fuels. Of those ninety-seven quads of total primary energy, 40 percent (mostly from coal, natural gas, and nuclear energy) goes to produce electricity, which of course is then used throughout the economy, 27 percent (almost all of it petroleum) goes directly to transportation, and 22 percent goes to industry and agriculture (divided about equally between petroleum and natural gas). The remaining 11 percent goes to the household and commercial sectors.

Fossil fuels provide services other than energy. About 9 percent of total primary fossil fuel use is used as industrial feedstocks, not as energy: fertilizer, pesticides, pharmaceuticals, plastics, textiles and artificial leathers, tires, asphalt, lubricants, and waxes—many of the things we rely on. Roughly 28 percent of those feedstocks come from petroleum, 24 percent from natural gas liquids, 11 percent from natural gas itself, 9 percent from coal, and the remaining 28 percent from cellulose materials such as wood scraps, sawdust, other byproducts of the lumber industry, cane sugar bagasse, and paper mill pulp.[4] No discussion of the role of fossil fuels in modern economies can afford to ignore their role as feedstocks.

THE FUTURE OF OIL AND GAS

What Energy Transition? The experts say that, so far, the world has consumed less than half, and perhaps less than one-third, of recoverable petroleum resources. Why are they so worried about running out of oil? The answer lies in the astonishing growth of the enterprise. The oil era really got underway only about 1940, and yet already—because of the speed at which consumption has grown—we can foresee the end of the petroleum era and the economic system that has grown from it. For the

United States, the domestic game is about over. Our crude oil production has been declining at an accelerating rate for thirty years. We now import 62 percent of our crude oil and, with less than 5 percent of the world's population we consume one-quarter of world production. We can eke out a few decades of dependence on oil and gas only if we can import it.

Worldwide discoveries of new oil fields peaked over forty years ago, despite intensified and increasingly sophisticated exploration efforts and extraction techniques. Non-OPEC production has probably peaked, and worldwide production is expected to peak very soon.

The U.S. Geological Survey (USGS) publishes one of the more optimistic estimates of oil and natural gas resources. Their estimates are summarized in Table 1 below.[5]

Table 1. USGS world summary estimates of conventional petroleum and gas resources, 50% confidence level.

[BBOE, billions of barrels of oil equivalent. Six thousand cubic feet of gas = one barrel of oil equivalent. Natural gas liquids included in petroleum figures.]

	OIL	GAS	
	Billion Barrels	Trillion Cubic Feet	BBOE
World (excluding U.S)			
Undiscovered conventional	796	4,333	722
Reserve growth (conventional)	654	3,305	551
Remaining reserves	927	4,621	770
Cumulative production	546	898	150
Total	2,923	13,157	2,193
United States			
Undiscovered conventional	83	527	88
Reserve growth (conventional)	76	355	59
Remaining reserves	32	172	29
Cumulative production	171	854	142
Total	362	1,908	318
World Total	**3,285**	**15,065**	**2,511**

Table 1 takes a bit of reading. Let me focus on the oil figures. To begin with, the figures for "undiscovered conventional" resources and "reserve growth" represent USGS' 50 percent confidence level estimates. If you asked, "How much of that oil are you *really* confident about?" the 3285 figure would go down to 2452 billion barrels. If you then asked, " . . . and that's what the world has left?" the answer would be, "Well, no; 717 billion barrels have already been consumed. We would bet 19:1 that there are more than 1735 billion barrels left, and we would bet 50:50 that there are 2568 billion barrels left. There may be 3343 billion barrels left, but we would bet 19:1 against it."

"Reserve growth" is a concept recently developed by the USGS. It is their guess as to how much more oil can be obtained by expanding existing known fields and applying new technologies to them. Growth enthusiasts greet every news report about the increasing efficiency of oil extraction, as if it vitiated old estimates of the limits of the resource. What they ignore is that allowance has been made for such improvements in the "reserve growth" estimate. And there is no assurance that those expectations will be borne out. Royal Dutch/Shell recently reduced its estimated reserves by 20 percent. That downsizing is ominous because it reflects a failure of a new technology, horizontal drilling.[6]

Why this statistical exercise? It shows that the USGS—one of the more optimistic players—is really confident only of about 1.7 trillion barrels of conventional oil remaining on Earth, with the odds diminishing to an even bet at less than 2.6 trillion.

How big is 1.7 trillion barrels? There are various ways of trying to make such stupendous numbers more concrete. One way is to divide the number by current annual consumption to arrive at a figure for a "resource/consumption ratio" or "years of consumption." Dividing the 1735 billion barrel estimate above by current annual worldwide consumption (29bb per year) gives a figure of sixty years. But this calculation is flawed. It assumes that world oil consumption will stay constant, which it won't.

The Energy Information Administration (EIA) expects it to grow 1.9 percent per year from 2001–2025.[7] At that growth rate, the resource life shortens to forty-one years.

That is still an unlikely scenario. Production will peak and then move down unpredictably, perhaps for a century. The important question really is: When will it peak? From that date on, rising demand will pursue a diminishing supply—and that is a recipe for intense competition and rising prices. The USGS does not put a date on the peak, but we can derive one by applying the statisticians' bell curve to the USGS world resource estimates. (That approach assumes that the peak will come when half the ultimate recoverable resource has been extracted.) The peak will come in 2015 if we use the USGS 95 percent probability estimate of 1735 bb and the 1.9 percent growth rate. The USGS 50 percent probability estimate would move it back to 2025.

Other less optimistic experts believe that the USGS overstates undiscovered oil and prospective reserve growth. Using a country-by-country bell curve analysis, they put remaining world oil resources at about one trillion barrels.[8] They calculate a peak before 2010, probably by 2007. The difference of opinion itself shows that this is hardly a precise exercise, despite the spurious four-digit precision of the USGS numbers. Nevertheless, the differences simply move the peak a few years one way or the other.

In Figure 1, I projected U.S. oil consumption declining from 25 percent of world production now to 12 percent in 2020 and thereafter (given our declining production, the rising competition on the world market, and our balance of payments deficit), using the Duncan-Youngquist projection of world production. And although it is a risk-laden world, I assumed no supply interruptions.

The analysis of natural gas is parallel to that of oil, though gas resources are harder to predict. Natural gas is a replacement for oil in many applications but a temporary one, since gas production will probably peak shortly after the oil peak. The USGS estimates for both fuels are similar

(Table 1), as are the resource/consumption ratios. The United States is fast running out of natural gas. Petroleum can be moved around the world cheaply by tanker, but moving gas by sea requires a cumbersome and costly process of liquefying the gas, shipping it by special tanker, and reconverting it to gas.

In Figure 1, I made some conjectural assumptions: a continuation of the slow decline in U.S. production, accelerating somewhat over the decades; and imports from overseas rising to replace diminishing imports from Canada until 2040 and then declining with the swift decline of world production.[9]

Intensifying Competition. Economists will try to adjust to the tightening world demand/supply equation and the entry into the market of new players such as China (where the automobile era is just taking off, and energy consumption—mostly petroleum—for transportation is projected to grow an astonishing 5.3 percent per year through 2025.[10] That would be a quadrupling by 2025, which requires a very optimistic view of availability.). Prices of oil and gas have already been rising, but those increases are negligible compared to those we may anticipate after the oil peak.

How far will prices rise? Nobody knows. Oil producers will try to maximize production from aging fields, which itself will drive up the price. Secondary price increases will occur in energy-intensive industries, as they pass on higher energy prices. Demand itself is unpredictable. Consumers will face a sharp decline in their standard of living as they absorb the higher price of energy, and this in turn will affect their purchasing power and propensity to consume. The monetary authorities—the Alan Greenspans—of the industrialized nations will face a juggling act much more difficult than anything they now have to deal with. The strains could precipitate a collapse of world fiscal and trading systems. The collapse would be most immediate and disruptive in the industrialized world and in the cities of the developing countries. It would have less effect on the rural folk in the less developed countries (LDCs)—who still constitute 60 percent of LDC population. They are already at or close to subsistence levels; but their countries would lose a part of their export markets and such food and aid as they have gotten from abroad.

The competition for energy resources may well spell the end of free trade. The United States is still betting that under the banner of free trade it will be able to buy the oil and natural gas that it wants. That may not work. In a few decades, when coal is king, the United States will have the largest endowment. Will we share it? With memories of the suspension of soybean exports by the United States when there was a poor crop, and similar European behavior when they had a poor grain crop, I am dubious. Welcome to the new era.

Few political leaders seem to recognize that the decline of petroleum is a new and fundamental issue. The U.S. Government seems to be mesmerized or in denial, and state and local governments continue to plan for growth and more traffic as though there were no energy crisis ahead. We may stay in that state of mind and stumble dumbly into disaster. More likely, we will search for every possible source of energy—but without addressing the population growth that drives the problem. There are several policy choices and energy sources to stretch out the crisis. I will itemize them below.

THE SEARCH FOR ALTERNATIVES

Conservation. At the next crisis, there will be strong pressures to conserve energy and use fuel more efficiently, and laws and tax policies should be changed to encourage it. The transportation sector is particularly vulnerable. SUVs will be a dying breed, and industrialists will become much less tolerant of energy inefficiency. There is very little to criticize in both those developments—though automobile manufacturers will lose their most profitable market. Climate will be under less human stress, and the environment will benefit. The United States has room for substantial savings. Look at Europe and Japan, which use about half as much energy per capita as we do, without suffering deprivation.

Unconventional Oil. There are immense deposits of bitumen in tar sands in Canada and unconventional "Orinoco extra heavy oils" (asphalts) in Venezuela. How big? The World Energy Council (WEC) estimates that there are 3.6 trillion barrels of oil in place in those two countries. That is more than the remaining conventional oil resources, but there will be a point at which it takes more energy to mine and convert those resources than they will produce. The WEC is less sanguine about those resources than the "oil in place" figure would suggest. It puts proven reserves at just 46 billion barrels, with another 193 billion barrels in probable reserves. They are being exploited now. The Canadian fields produce 500,000 barrels per day, and production is being upped to one million barrels. It is said to be profitable at today's oil prices, even though extensive processing is needed to convert it to conventional oil. Petroleum geologist Walter Youngquist observes that there would be "enormous" problems in scaling unconventional oil production up to five million bpd. If that rate can be achieved, the proven and probable reserves would last over a century, adding over 2 percent to present world oil production, which is the single biggest boost in sight for hydrocarbons.[11]

In Figure 1, I assumed that the United States could purchase 2 million of those 5 million barrels until about 2085 and then taper off slightly as the "probable reserves" pass their peak.

Problem: Tar sands and extra heavy oils are loaded with pollutants such as heavy metals and in processing they release large quantities of carbon dioxide, which will force climate warming.

Oil shales introduce the question: When is a resource an economic source of energy? Hydrocarbon-bearing rocks, worldwide, may well contain tens of trillions of tons of kerogen, which is related to oil but must be extensively processed to become oil. It must be mined, subjected to intense heat, and reconfigured to add another hydrogen molecule (which requires a great deal of water), and there are immense tailing piles to be managed. There was an oil shale bubble in northwestern Colorado in the

1970s, where several major oil companies lost literally billions of dollars. Since then, the general wisdom has been that oil shale is an impossibly expensive way to make petroleum (although a few small plants are operating in Australia). The richer shales can, however, be mined and simply shoveled into a boiler and used to generate electricity, a process already in use in Estonia.[12] How much of this can be done is anybody's guess; it is too far from realization to offer an estimate.

Problem: The volume of wastes is huge. Like popcorn, the material expands as it is processed. And the air pollution and carbon dioxide releases have yet to be quantified.

Coal. Coal was the first of the fossil fuels, the dirtiest, the slowest-growing, and it will be the last to go. Estimates of the resource are based on different assumptions and definitions. The DOE/EIA puts world reserves at 1018 billion tons and U.S. "demonstrated coal reserves" ("proven" plus "indicated") at 455 billion metric tons.[13] The World Energy Council (WEC) puts proven world reserves at 984 billion tons and proven U.S. reserves at 250 billion tons.[14]

Americans can draw some comfort from the WEC estimate. It assigns the proven reserves this way: United States 25 percent, Russia 16 percent, China 12 percent, India 9 percent, Australia 8 percent, Germany 7 percent, and South Africa 5 percent. The other half of the world's population will have to import coal or get by on the remaining 18 percent.

That distribution will be a major determinant of different nations' long-term economic prospects. Coal is a major source of electricity, and it can be liquified (at a loss of half the energy) into a substitute for petroleum. Germany made synthetic gasoline in World War II, and a commercial syn-fuel operation is running in South Africa now.

Coal reserves of 984 billion tons are equivalent in energy value to 4.6 trillion barrels of oil. That is more than the USGS estimate of 4.5 trillion barrels of oil equivalent (tboe) of remaining oil and gas in Table 1, and the figure for coal does not include the undiscovered exploitable resources, whatever they may be. The comparison suggests that coal is a huge resource, but not an infinite one, and it is costlier to extract than oil. The WEC "reserve/production ratio" for coal reserves for six of the seven leading countries (except China) comes out to over 200 years' supply, but that calculation is nearly meaningless, both for the reasons I cited in the petroleum discussion and because the demand for coal will skyrocket. It provides just 27 percent of world fossil fuel consumption now;[15] it will be called on to provide much of the other 73 percent as oil and gas wind down.

Thus, coal can play a pivotal role in the energy transition, providing both energy and chemical feedstocks until we can get our house in order and learn to live within a renewable resource economy (Chapter 3). If the numbers are right, the United States will still have diminishing production until after 2150, but the energy transition will be dangerous and difficult; and we would be wiser to cut back on coal production when we can, thus preserving it as a backstop and chemical feedstock for a longer time.

About half the reserves consist of anthracite and bituminous coal and nearly half of sub-bituminous coal and lignite. Coal is bulky and dirty in just about every way, and lignite is the worst.

And that leads us to the central question about coal. It can help to soften the transition as oil and gas production decline, but the environmental costs could be immense. The WEC 2001 report offers the table on the following page.

Moreover, the WEC estimates that, over the production/consumption cycle, coal emits 50 percent more carbon than natural gas and 25 percent more than petroleum, per unit of energy, and thus is a worse source of climate warming.

For those in doubt about coal's noxiousness, one has only to remember stories of the killing smog in London, the product of countless little coal

Table 2. Comparison of Air Pollution from Different Fuels
Kg of Emissions per TeraJoule of Energy

	Natural Gas	Oil	Coal
Nitrogen Oxides	43	142	359
Sulfur Dioxide	0.3	430	731
Particulates	2	36	1333

Source: Worled Energy Review 2001

fireplaces, or breathe the acrid smoke that envelopes Chinese cities as housewives light their coal-and-mud briquettes to make supper.

Reliance on coal and oil sands is a dangerous course. Aside from the greenhouse gases and atmospheric pollution, coal mining always disturbs the land. Strip mining, which sometimes involves cutting the tops off mountains, is the most destructive, and efforts to enforce restoration of the land have had mixed success. Coal mining uses and degrades valuable water resources that are particularly scarce in the U.S. West. But the atmospheric pollution can be markedly reduced, and the byproducts may be useful as feedstock, through a process with the forbidding title of Integrated Gasification Combined Cycle (IGCC).

In the IGCC process, the coal is gasified, the impurities removed, and it can then be burned as a gas for electric power generation or converted to a liquid energy substitute for petroleum. Among various experiments with IGCC, the most notable was one conducted by a consortium of energy producers and users at Cool Water, near Barstow, California, in the 1980s. It could use high-sulfur coal and even sell the sulfur at a profit. That experimental plant was not competitive at the prices of that period. It was dismantled and its components reassembled in Kansas as a way to convert coal into urea fertilizer. Interestingly, that plant is thriving now, when many factories that use natural gas as a feedstock are in trouble because of the rising price of gas.[16]

At least two new, larger IGCC projects are underway in this country: one in Florida and one in Indiana sponsored by the DOE Clean Coal Demonstration Program.

Coal consuming countries face a momentous choice: Will they burn coal the way they have been doing, and set environmental disasters in motion, or will they go to IGCC and pay a higher price for clean power and feedstock from coal?

That still doesn't solve the biggest problem. The IGCC process does not address the climate problem. It can use some carbon dioxide as a chemical feedstock, but not enough to significantly reduce the overall carbon emissions from burning coal. If we are to avoid compounding the human effect on climate (discussed later), other ways must be found to achieve that reduction.

As oil and gas taper off, the overall rate of global warming will depend largely on whether we can learn to sequester the carbon dioxide released by the use of coal and oil sands. That hasn't happened. The present proposals call for injecting it as a gas into old mines or wells. I have yet to see an inventory of existing subterranean spaces, and CO_2 takes up much more space than coal, so every ton of coal burned will generate hundreds of thousands of cubic feet of CO_2.

The proposal may also be a very dangerous one. As a gas, the CO_2 will seek to escape, particularly if it is under pressure. If it does escape, sequestration will fail. Moreover, CO_2 can be a silent killer. This has happened. Natural CO_2 seeps up through two "killer lakes" in Africa. Such releases a few years ago killed some 5,000 lakeside dwellers. CO_2 is odorless and heavier than air; it simply suffocated the unsuspecting villagers.

It would be better if the CO2 could be incorporated into some inert solid, but CO2 is one of the most stable of molecules, and no proposal has been forthcoming that would immobilize it in a solid form, economically and without major environmental costs.

Problems: (1) Sub-bituminous and lignite coal apparently cannot be used in IGCC plants, which means the best that can be done is to scrub them, which itself creates huge quantities of used limestone slurry. (2) The industrial nations have yet to go for the expense of IGCC. Developing countries such as China and India seem even less likely to do so, because it is expensive. (3) Sequestration by whatever means is a monumental task. What do you do with some 16 billion tons of CO2 a year? (For a sense of the scale, consider that that is eight times as much tonnage as all the world's annual grain production, and it is vastly larger because it is a gas.) DOE has a target of doing it, eventually, for $10 per ton, but that is only a target, set by a protagonist of sequestration. By another estimate (or rather, guess), it would cost $80–$100 per ton "assuming that those technologies can be developed"[17] —and that too is simply a speculation. It may not be doable.

In Figure 1, I increased U.S. coal production 2 percent per year until 2070. (Remember, Figure 1 assumes no policy change, and our present policy is growth.) By then, we will have reached the midpoint of known reserves and, following the bell curve, I project declining future numbers. This projection may be low if some of the USGS "indicated reserves" prove out; it could be high if the environmental disruption and climate damage become overwhelming. Current net exports are less than 4 percent of production. I assumed they would continue but not increase in an era marked by energy stringency. By the latter part of this century, the rest of the world will be in a much more desperate energy crunch than the United States, and that assumption about exports may not hold true if they can find ways to pressure us to increase them.

After coal, we move into energy sources that cause less air pollution and do not contribute to climate warming, but they are more problematic as a way of replacing oil and gas.

Nuclear Energy. Nuclear fission is an established source of electric power. Nuclear power plants—over 400 of them—exist in many countries. France generates over 70 percent of its electricity with nuclear power. The United States has about one-fourth of the plants and one-fourth of world production, but we stopped building new plants a generation ago.

The resistances to nuclear power have limited its introduction elsewhere. Those resistances arise from (1) the fear that rogue states may divert uranium from nuclear power production to make nuclear weapons, and (2) the concern that radioactivity from high-level nuclear wastes will escape storage and pollute the environment, or that widespread nuclear pollution may result from accidents such as that at Chernobyl. The threat is nearly perpetual, given the radioactive half-life of 10,000 or even 300,000 years of some of those materials. Nuclear proponents counter that nuclear energy has been remarkably safe. Even the extent of the damage from Chernobyl is hotly contested, and the safety of a proposed disposal site in Nevada is a major political issue. The debate has polarized, and it is very hard to arrive at an objective judgment of the seriousness of the threats, partly because to a unique degree they depend on human behavior.

As energy shortages develop, however, it is a good bet that countries will turn increasingly toward nuclear power. For the uncommitted, the reasoning will be that, whatever its faults, nuclear power is better than no power. Let us hope that, in exchange for that acquiescence, a world system is created that can manage the threats.

That acceptance would not obviate another problem. Fission energy is itself limited, because uranium resources are finite. The International Energy Agency (IEA) estimates "reasonably assured resources" of uranium (at $130/kg or less) as 3 million tons, with "estimated additional resources" of nearly another million tons. Annual consumption has been flat at about 62,000 tons. If it stays flat, that would provide enough uranium for forty-nine to sixty-five years. It is estimated that U.S. uranium resources would last for thirty-five to fifty-eight years even if the country were to quadruple its nuclear electricity production.[18] (These numbers may go up with new discoveries.)

In Figure 1, I elected to project U.S. nuclear power production as flat, assuming that new plants will replace those being retired. If we did indeed multiply the production, it would be so short a future that it would be a questionable investment. My projection is conservative; we might increase the capacity, particularly if we go in for reprocessing, if some new resources turn up, or if rising uranium prices lead to increases in the reserves.

The WEC points out that the limited nature of the resource has been partly obscured in recent years because some 40 percent of the uranium needed for power generation was acquired from existing stockpiles and the conversion of nuclear weapon stocks.

The French extend the horizon somewhat. They own the Cogema mines in Canada, the largest uranium producers in the world. They also reprocess the spent fuel rods. However, the rods are reprocessed only two or three times, after which the different radioactive byproducts become so nasty that the French put them into vitrified storage and start anew.[19] The French and Japanese, both without fossil fuels, experimented with sodium-moderated breeder reactors, but that is indeed a dangerous game. Liquid sodium is a superb heat sink but a very tricky material to handle, and as of this writing, both experiments were on hold.

Australia has 20 percent of the exploitable uranium resources, Kazakhstan 18 percent, the United States 11 percent, Canada 10 percent, and South Africa 9 percent. The remaining 32 percent is widely distributed, so an OPEC-style oligopoly is some distance off.

Problems: Fission produces only electricity, not concentrated mobile energy or chemical feedstocks. That limits its use to less than 40 percent of the energy market; and like fossil fuels it has a limited time horizon. And if our demand for energy leads us to go with fission, we are making a Faustian trade-off of energy for ourselves, in exchange for our descendants' having to live with the threat of radioactivity escaping from confinement.

ENERGY, THE ENVIRONMENT AND GLOBAL WARMING

Let me go back to the earlier discussion about the petroleum peak. Turn those calculations around: Humankind has burned less than half the petroleum that we expect to burn. The environmental problems it has created will double, or more. We have used only 12 percent of the estimated, ultimately recoverable gas, so we have eight times as much pollution yet to come. If we burn the estimated reserves of coal without extracting the pollutants, the total nitrogen oxide emissions will be about five times as much as we may expect from oil (Tables 1 and 2). About 3.4 times more sulfur oxides will come from that coal than from petroleum, two and a half times more carbon, and about sixty times as many particulates. If we knew the amounts, we could add the damage from heavy oils. The numbers suggest a dire conclusion: We are in an overshoot mode, not just because the fossil fuels are running down, but because of what they are doing to the environment and the climate. By extracting carbon, nitrogen, and sulfur from the lithosphere and injecting them into the atmosphere and biosphere, we are embarked upon a fundamental alteration of our habitat. People seem to have become blasé about that prospect.

We have learned something about controlling air pollutants. The lesson is yet to be applied in the less developed countries. Even in the United States, the EPA reported in 2002 that aggregate atmospheric pollutant emissions are again on the rise. And we haven't mastered the CO_2 emissions.

Atmospheric CO_2 began to rise in the 1700s, as coal came into general use for power and home heating. It accelerated with the petroleum/gas era. All fossil fuels generate atmospheric CO_2 and drive climate warming.

The anticipated consequences have been discussed at length elsewhere: multiple health hazards; hotter and drier tropics, and less food production where it is most needed; changes in forest cover, and a loss of forests if climate warming moves faster then tree species can migrate; heat waves; increasingly erratic droughts and storm cycles; the alteration of stream flows; a warming and rising ocean. Of that literature, I will focus on only one issue: the effect on sea level.

The sea is in farther retreat down the continental shelves than at any time in the past 200 million years, except for the glacial periods of the past million years.[20] However, sea level has risen in recent decades, and the International Panel on Climate Change (IPCC) estimates that it will rise about twenty inches in this century. That very small rise can drive shorelines miles inward on a very gently sloping shelf such as in Florida, and it will expose new areas to storm surges. The recent rise in sea level is the result mostly of thermal expansion. The shelf ice in Antarctica has been breaking up, and the sea ice in the Arctic has thinned by almost one-half in the past half century. Those events have not affected sea level because that was floating ice and, obeying Archimedes' law, its melting did not change sea level. Mountain glaciers have been retreating worldwide, but the melting glaciers' effect on sea level will become noticeable only as the Antarctic and Greenland ice caps melt, which is starting to happen.

The IPCC analyses are confined mostly to effects within this century, although it has pointed out that even the present level of anthropogenic greenhouse gases will affect climate for centuries to come. Unpredictably, and probably over a long timeframe, the sea may recover some of the coastal plains that it has given up. Melting of the ice caps could raise sea level about 100 meters. (It was apparently higher than that in the late Cretaceous.) Over half the U.S. population lives in coastal counties. A substantial share of the Earth's population lives in potentially threatened zones below 100 meters elevation. Unlike a recent movie about global warming, the change will take a long time, and people will have a chance to retreat from inundated areas or those swept by storm surges. But where would they go? On a fully occupied Earth, the upland residents would resist the movement.

Turning this threat around is not an easy task. In 1995, the IPCC calculated that the only way to hold the climate impact of human activity at its present level would be to reduce carbon emissions by 50 to 70 percent right away, "and more later." Nothing on the political agenda even begins to address that challenge.

What can we do about the environmental threats and expectation of climate warming? The only reasonable ways are to (1) minimize the threats we can manage, and (2) slow down the rate of emissions by reducing demand. That second policy is a powerful argument for a deliberate policy of reversing human population growth. It offers the hope of escaping the penalties of fossil fuels, even though there will be tremendous problems of adjustment.

RENEWABLES

One way to minimize the threats would be to go to less polluting renewable sources. That is, by and large, a long-term process demanding many fundamental changes, and will be discussed in Chapter 3. Let me mention here the renewables that play a role in the present energy budget.

Hydropower is a known quantity. It already provides about 3 percent of U.S. energy, but it is unlikely to go higher. The best sites have been occupied, and some old dams are being retired because of their impact on salmon runs. The world outlook is not much better. The WEC estimates that world undeveloped hydro potential is twice the current capacity, but population density in the less developed countries generally means that the valleys behind the proposed dams are thickly occupied. As China is finding out, new hydropower sites involve some painful tradeoffs between the costs of displacement and the power, flood control, and irrigation gains that the dams make possible. The Three Gorges Dam on the Yangtze required the removal of more than a million farmers. It is a high cost source of power.

Hydropower is classified as renewable, but the term is relative. The reservoirs behind the dams will silt up—in something like a century or two in eroding areas—and the hydropower will simply become run-of-river (i.e., utilizing the river's variable flow but with no storage capacity).

Climate change enters here. The models agree that global warming will result in more concentrated and erratic storm systems and faster runoff from winter alpine snowpacks, and that seems to be happening now. That is bad news for hydroelectric generation: more erratic streamflows, more erosion into reservoirs,

Problem: Hydropower will be a minor source of new energy at best. Little or no additional capacity is likely in the industrial world, somewhat more in the less developed countries. If total hydroelectric production were doubled (a very optimistic assumption), it would meet less than 7 percent of present world electric power needs and progressively less in the future.

Biomass is already a significant source of energy, worldwide. As we shall see, its potential for expansion is presently limited in the U.S. and particularly elsewhere by the competition for the land on which we must produce energy for biomass energy. However, it needs to be encouraged and developed now, along with Wind, Photovoltaics, and the more speculative future sources, to play its eventual role. The transitional fuels I have inventoried are simply that. They give us some time to move beyond them, but it would be foolhardy to wait until they are gone before we stop and ask: Now what?

In Figure 1, I have held U.S. renewable energy at its present level, since significant increases will require far more dramatic actions than those presently in sight and are the subject of Chapter 3.

PREPARING FOR THE POST-FOSSIL FUEL ERA

The Slowly Gathering Storm. That discussion of energy options suggests that for most of this century, dwindling oil and gas resources can be augmented by unconventional hydrocarbons, coal, and nuclear fission; but those alternatives are dangerous, limited, and/or expensive (particularly if we try to avoid escalating environmental damage), and only coal has the potential to accommodate the projected growth of U.S. population until about 2070.

Beyond those transitional forms of energy, there looms an immense qualitative change as the fossil era comes to an end, and as we move from competition for a diminishing resource to the need to find new energy sources to replace those on which modern economies have been built.

Most of us use the word "transitional" to speak of the pending energy shift. The word implies a bridge from one state to another. Usually, when building a bridge, the engineers understand the nature of the terrain at both ends. In the energy transition, however, we are proposing to build a bridge into a void. We don't know what is at the other end. That void will be the subject of Chapter 3 of this paper.

The Choices Before Us. In that uncertain condition, there are several strategies available to the United States and, less certainly, to other countries.

- Prolong the transition as long as we can. Energy consumption can be lowered by reorganizing our living patterns, by energy-efficient business and manufacturing processes, and by promoting public transportation, more house insulation, and passive solar house heating. Some of these changes will result from rising energy prices. Others may require a willingness to deliberately substitute long-term energy savings for short-term convenience. That willingness has been suicidal in modern American politics; it will come about only if it is demanded by an environmentally literate public.

- Minimize the impact that the dirty transitional fuels will have on climate change and the environment. We must find a way to sequester the carbon emissions from fossil fuels, including coal and heavy oil. If we cannot, we are heading straight into a worldwide environmental disaster.

- Prepare for an unknown future on the best possible terms. Political leaders everywhere will need to give up their fixation on growth as a panacea for economic ills, because growth abbreviates the transition and rushes us into a future for which we are not prepared. Planners will need to abandon their enthusiasm for energy-intensive economic models such as suburban living, skyscrapers, superhighways and automobiles, and airplanes. Common folk must be prepared to live a simpler life as rising energy costs erode their real incomes—and businesses should be ready to provide workable alternative transportation and living arrangements to which they can resort. Scientists should be mobilized to try to clear the murk ahead and describe the alternative energy systems that may work, so that we may begin serious work on building that bridge.

- Above all, seek the solution on the demand side. The United States must stop and reverse its population growth so as to match the decline of conventional energy with declining demand, and to free more resources for the investments that lie ahead in converting to renewable energy. We must come to a consensus and start the move now if—as I think Chapter 3 will document—we learn that the post-fossil energy resources will not support the population we have, at a level above penury—or if we learn that we cannot limit the damage from burning coal. We must learn to put such calculations ahead of the parochial agendas that have stood in the way of a population policy, such as the unwillingness to address the mass immigration that presently drives U.S. population growth, or the political posturing that has crippled U.S. programs to help third world countries stop their population growth. We must, in short, embark on a new agenda that seems hopelessly out of touch with present political realities but that may become more realistic as we recognize the extraordinary changes that must be accommodated as we move toward the end of fossil fuels.

Chapter 3.
Twilight or Dawn?

Chapter 2 addressed the period of decline of fossil fuels, lasting perhaps through much of this century, and concluded that the fossil fuel era has been a brief spike that generated an unsustainable growth in population and, in the industrial world, consumption. We are entering an age of overshoot. Humankind may have time for an orderly withdrawal if we have the wisdom to face the change squarely and to unlearn the faith in growth that has grown up in the fossil fuel era. There will be disasters if we do not, and later wisdom may come to see the fossil fuel era as the most intense human disturbance of our natural support systems in history.

Chapter 3 will look toward a much more uncertain future beyond fossil fuels. With the judicious employment of the technologies we have learned—and with a bit of luck—we may be able to create a more harmonious balance with the rest of the biosphere, but only at substantially lower population levels and less consumptive habits.

***

Many studies have undertaken to describe how benign renewables might replace fossil fuels. A few of them have noted that population growth makes the task more difficult. Almost none of them turn the issue around and make the point that only a smaller population will make it possible. That is the central point I hope to make.

Once we have left the comforting but dangerous shelter of fossil fuels, we will necessarily turn to biomass, wind, direct solar energy, and some more exotic sources. Biomass is traditional, and it is the most versatile form, but it will be a strictly limited resource unless human populations have grown much smaller, and the other renewables in sight can fill only a small part of the energy void, especially if demand does not decline dramatically.

In this speculative world, estimates of one or another source vary by a factor of ten or more. Generally, the high estimates cite a theoretical figure for availability, without considering the costs, the relevance to actual human needs, or the question: Can the source yield a net energy gain over the energy put into obtaining it? The low-end figures come from those who study those limits. I am inclined to go with the skeptics. I will cite some numbers that seem reasonable, but will leave it to the footnotes to guide readers to more detailed studies.

BIOMASS, THE BEST HOPE?

Biomass (particularly fuelwood and charcoal) was and is an important source of energy. At its simplest, biomass is simply fuelwood, which was the major form of energy used in the United States through1880. Wood and grass are still the principal energy sources in rural areas in the less developed countries. We do not really know how much biomass is used for energy. The World Energy Council (WEC) cites estimates of 5 to 9 percent of total world energy.[1] The official figure for the United States is 3 percent (which is probably low).[2]

A ton of wood generates about 60 percent as much energy as a ton of coal, and grasses 50 percent. They have been used for fueling steam engines and for industrial power. In Brazil, 6 million tons of charcoal are used each year in heavy industry such as steel making. Biomass can be liquified into alcohol and methanol, and it has been used as a motor fuel. Oilseeds and soybeans are easily converted to "biodiesel." Bagasse, the residue from sugar cane processing, constitutes three-fourths of the weight of the cane. It is used for power and fuel for cane mill operations. In Brazil, India, Mauritius, and Thailand, the surplus electricity is sold to the grid.[3] As a chemical feedstock, biomass can substitute for fossil fuels. Properly utilized, it generates very little pollution and does not cause climate warming.

Thus, biomass is a flexible and benign energy source uniquely capable of filling the roles that fossil energy has filled. But there is a catch: It competes for land and water with food, fiber, and lumber production. There will even be competition from new uses of land to raise chemical feedstocks and pharmaceuticals. And finally, climatologists are proposing the massive expansion of forests as a temporary way of reducing carbon emissions. The worldwide competition for land and water is intense and it is getting worse, because arable land per capita is diminishing everywhere.

A significant early worldwide gain in biomass energy is unlikely, unless the rich outbid the poor and produce commercial energy on land that is already producing food. A long-term gain awaits a dramatic decline in the overall pressure on land and water resources—through a remarkable increase in crop and forest yields or, more realistically, by a turnaround in population growth.

Therefore, the prospects for biomass energy will rise or fall with demographics. With a small enough population, biomass could fill much of the energy gap.

The United States is in a relatively good position, for a while. We still have four times as much arable land per capita as the less developed countries, and three times as much as Europe, but our arable land per capita is declining rapidly. In another generation or two, population growth will eat up the grains we now export, unless we find a way to raise yields (which have stagnated) or change our eating habits; and after that comes the question: How do we feed a growing population? The competition for land is coming home to the United States, and that does not look promising for biomass energy.

There are some practical opportunities in the United States. The commercial pine plantations in the Southeastern United States produce lumber, paper, and pulp. They can be harvested for energy if energy prices rise enough to compete with them. But there is a long-term threat: Models of global warming agree that the Southeastern U.S. will be getting hotter and drier, and probably turning from piney woods into savannah—and that is not a good portent for the pine plantations. This is an argument not usually heard for stopping anthropogenic climate warming.

In the American West, we have learned from recent disastrous forest fires that softwood stands have become much too dense, as a result of bad management for two centuries. As an interim measure, those stands should be thinned and the biomass harvested.

Some biomass energy could be gained from unused crop residues, forestry residues (scrap lumber and sawdust), and urban wastes that are presently wasted because we do not have the management systems and market demand to put them to use. In the European Union, "pollution taxes" are being used to penalize fossil fuels. We could use the same approach to promote the more systematic use of renewables such as biomass. That could provide the foundation of a strategy for the transition into new energy sources.

There are limits. First, biomass is a diffuse energy source. Until petroleum began to run down, we simply had to drill a hole and let a highly concentrated form of energy flow out. No more. Biomass usually converts about 0.1 percent of the solar energy striking plants into usable energy through photosynthesis. Unless the biomass is a byproduct, like bagasse, it is expensive to harvest. It must usually be fertilized to justify the cost of harvesting. And some of the biomass will need to be converted into motor fuels to harvest and transport the biomass itself, or recycled into fertilizer to raise more biomass (because commercial fertilizer is now made from fossil fuels, which will not be available after the energy transition). Less energy-dense stands do not pay back the energy used to raise, harvest, and process them, unless they are harvested, traditionally, by peasants who have no opportunity costs for their labor. (The U.S. Government subsidy of ethanol alcohol from corn—the grain, not the stalks—was dictated by politics rather than economics. Most experts believe that it produces less energy than went into making it.[4])

Second, biomass is not necessarily benign. Burning it in traditional faulty stoves or fireplaces can make it a killer, or at least a potent health problem. (I remember visiting Himalayan peasant huts and being driven out by the wood smoke.) We hardly want to go back to the good old days. And, unlike oil or natural gas, biomass harvesting can strip the tree cover and denude the land, as growing populations attempt to supply their energy needs. That is a pretty fair description of the poorer countries now. Fuelwood and grass are collected from marginal woodlands, hillsides, roadside and village trees, orchards, rubber plantations, and grasslands. It is hard work, often done by young women close to the economic margin. We don't want to perpetuate that scenario. It is hardly the prescription of the future that most of us have entertained, but it may necessarily still have a role because it is better than no energy. That again is a function of crowding. With a better population/resource ratio, there will be sufficient land and water to get the required biomass from less brutal collecting practices.

Third, the annual flow of sunlight imposes limits to biomass energy production. Fossil fuels are the biomass of earlier eons, and we have been burning them to support our energy habit—perhaps twice as fast as the Earth's total annual solar input via photosynthesis. In the future, that annual budget must support all the world's life, including our own needs for food, fiber, and biomass energy. Humans already use a substantial portion of total photosynthetic energy,[5] taking it from the other living creatures. We cannot use more biomass energy without imperiling the biosphere and ourselves, unless we sharply reduce our other demands on land and water resources. And, mathematically, the upper limit will be far below the present brief spurt of fossil energy production.

How far below? From the welter of estimates of biomass potential, let me analyze one that is near the middle of the pack—that 350 million hectares could produce 80 Exajoules (EJ) of energy per year, about one-fourth of current total world energy consumption.[6] Corn may yield about fifteen tons of biomass per hectare, which is consistent with that estimate. However, most farmland produces much less biomass. Since 350 million hectares is one-fourth of total world cropland, a more realistic figure for average biomass production would be eight tons or less, thus halving the gross energy estimate to forty EJ.

Even that estimate is unrealistic under present conditions because the world cannot afford to turn over a quarter of its diminishing cropland to biomass energy production; it needs that land for food and fiber production. We could get part or all of those 350 million hectares from tree plantations on presently underutilized forest land, but that would cut the energy production to much less than forty EJ because trees grow more slowly than grasses, and their biomass yield is unlikely to exceed three tons/hectare—especially since the land available will be marginal. Such land is unlikely to add much to the present biomass harvest—part of which already comes from peasants cutting wood on that land. It would take heavy fertilization and management to raise the production and even then, the harvest might not justify the energy inputs.

For the United States, I have pointed out that our situation is relatively good, but declining fast. Toward the end of the century, when we will need the biomass, anticipated population growth will have ended our food surplus and wiped out any prospects for biomass energy.

Demographics is the key, in the United States and elsewhere. A world of half the present 6.4 billion could release a quarter of arable land to biomass energy production, and still have 50 percent more arable land per capita for food production. That may not be enough in a world without fossil fuels for fertilizer production (see Agriculture below). If we halve the population again, to 1.6 billion (back to the 1900 level), there is ample land for food and energy production. We would be better off than in 1900 because we have learned better ways of using land and water.

The world got by on biomass for almost the entire span of human existence.

THE SPECULATIVE SOURCES

One can understand the near-desperation with which people look toward wind, solar, and hydrogen, given the limits on traditional fuels, but wishing is not necessarily enough.

Wind. To a point, wind is the next best hope. The technology is in place, and its cost (at the wind turbine) is in the neighborhood of 5¢/kWh (kilowatt hour), which is about one-third higher than conventional power in the United States. One regularly hears statements that there is enough wind in the United States to provide some multiple of our total energy needs, but they apparently are based on a rough theoretical calculation of all the wind energy in the United States, ignoring the question: What kind of wind?

The energy in wind varies with the cube of the wind speed. An eight meter per second (mps; i.e.,eighteen mph) wind has half again as much energy as a seven mps wind. Commercial wind energy becomes practicable only at about seven to eight mps. Only in limited locations does the wind blow regularly enough and hard enough to make wind turbines produce more energy than it takes to build and operate them. The American Wind Energy Association (AWEA) has made grandiose claims, but its practical target is much more modest: to supply wind power for 6 percent of U.S. electricity in 2020 (i.e., 2.4 percent of anticipated total energy needs).

And most grand claims for wind power ignore the question: What do you do when there is no wind? I will discuss the problem of intermittency later. One cannot predict very accurately when the wind will blow. Even if one could, it is unlikely to blow in synchrony with power needs. It becomes very difficult to incorporate erratic wind power into a conventional grid once the proportion of wind power passes 20 percent. Germany is approaching that level now, and Denmark has stopped adding wind power.

While fossil fuel energy is available, expansion beyond that line may not be economical. When fossil fuel is gone, wind may be the best of some bad choices. Utilities regularly maintain excess gas-fired capacity, which stands idle most of the time, to meet peak load demand. Wind may eventually help to meet that demand, or perhaps to power some form of energy storage. However, the shortage of good sites will probably limit our wind energy output.[7]

A shift to wind will be expensive. Wind turbines generate only about 20 to 30 percent of their rated capacity, because of erratic winds, while a coal-fired plant may operate at 90 percent of capacity. The installed costs of both types of plant are roughly similar, so the capital costs of wind energy will be three or four times that of the coal plant.

Wind power is relatively benign, environmentally. One problem is the noise, which can be heard for over a mile from the new large turbines. Europeans, accustomed to their tidy landscapes and population densities, find that a serious problem. In the United States, the best wind sites are offshore, in certain low mountain passes ("low" because air density decreases with altitude) and in the High Plains. In most such locations, there are not too many people to object to the aesthetics—if indeed a row of windmills on a lonely ridge in West Texas is thought to disfigure the landscape. Another problem is the slaughter of birds. With the modern enthusiasm for electronics, tens of thousands of towers are going up all over the United States (and we assume more elsewhere). The Audubon Society complains that millions of birds are being killed annually, and the addition of thousands of huge wind turbines to the landscape would multiply the problem. One can only hope, rather helplessly, that some way will be found to shoo the birds away from such structures. How long does natural selection take to develop birds that will dislike the towers?

Nevertheless, as conventional energy prices rise, wind seems a good bet for some of our power.

Direct Solar. Solar energy, because it is the object of so much enthusi-asm, is the subject of wildly varying expectations. I will thread my way through some of the more cautious speculation.

Photovoltaic energy is growing, but from a tiny base of less than 0.0006 of U.S. energy production. It is already useful as a niche source of energy, primarily for standalone applications needing very little electric power. It has been touted as the "energy of the future," but it may never replace most of the power that fossil fuels routinely put into the grid. Right now, subsidized by favorable state legislation, solar enthusiasts are reselling power to the grid at about $0.10 per kWh, three times the utilities' whole-sale price, and still the enthusiasts are losing money. A student of renewables, Ted Trainer, has estimated the cost of building and operating a solar plant in Australia over thirty- and twenty-five-year life cycles and found it thirty-three to forty-seven times as expensive as a coal plant, even before factoring in the high maintenance costs of solar power.[8] Another student, Andrew Ferguson, points out that even if enough direct solar capacity could be built to meet peak U.S. electricity demand (because peak demand and peak sunlight coincide), a perpetual annual investment of $110 billion each year would still provide only 24 percent of present U.S. total electricity needs because solar power functions intermittently.[9]

I would add that 24 percent of electrical demand is just 10 percent of total U.S. energy demand, and that his $5/watt capital cost estimate is five times that of wind power. And Ferguson is describing only the capital costs, not maintenance and cleaning (dust on the panels kills their effi-ciency; and solar works best in sunny and, therefore, dry and dusty, locations). He did not include the cost of the capital, distribution costs, and the energy losses associated with moving it to distant locations on the grid, or converting it to AC.

Advocates promise that costs will come down. Skeptics point out that most of the cost of a solar plant is the supporting structure, that there are few technological gains in sight there, and that the price will rise as energy costs drive up the price of the materials.

Photovoltaic power demands a lot of space. To stake out a maximum, the area of solar collectors needed to provide all present U.S. power needs (intermittently) would be 15,000 square kilometers[10] (5800 square miles). That figure should probably be quadrupled in order to spread out the panels to prevent them from shading each other, and to provide space for support structures and access lanes. That would monopolize an area almost one-fourth the size of New Mexico. The birds, lizards, and environmentalists will not be happy.

Photovoltaics may be more successful as part of a mixed energy future, generating "distributed power" such as electricity for household use in sunny climates, storing the energy in batteries for night time or cloudy weather. It will be very expensive electricity, and it won't run many air conditioners or heaters. Perhaps more important over the long term is passive solar energy—the use of sunlight to warm buildings by properly designing and placing them. From personal experience, I know the potential savings. This is a marginal role, however. Home heating is a tiny fraction of U.S. energy consumption, most housing is badly designed for passive solar, and it will take generations to replace the present housing stock with solar houses.

Solar thermal power, another form of direct solar, is designed for grid use. It uses sunlight to heat an oil or salt energy carrier, which is then used to generate steam with a conventional boiler. One such plant in California (with some conventional backup) produced power for the grid at a cost of 8 to 10 cents/kWh, over twice the price of conventional power, but went bankrupt. At one time, it was said to be the producer of 90 percent of the direct solar energy actually delivered to the grid, worldwide. Such a plant would need about ninety-two square kilometers (thirty-three square miles) to match the output of a conventional 1000 MW fossil fuel plant.[11] A coal plant takes only a few acres, but the difference narrows when one includes the coal mine and the dedicated railway line that brings coal to the plant, and solar thermal is much less disruptive to the environment. Despite its misadventures, the California solar thermal plant may show the way to a useful renewable electricity source. Expensive, but all power will be expensive.

Intermittency and the Storage Problem: the Hydrogen Dream. The bugbear of wind and solar energy is their erratic nature. Modern societies want reliable energy. Renewable energy enthusiasts propose that excess wind and solar generating capacity be used to store the energy for later use. The question is, how?

There have been various ingenious proposals: storing heat in salt; compressed air stored in abandoned mines; giant flywheels; and pumped storage, which is simply water pumped back up into a reservoir with excess energy when demand is slack, to be run through the hydroelectric turbines again. (The loss of energy in that transaction is justified by the price differential between peaking power and slack demand periods.) Pumped storage is practical and already in use in limited situations; the other proposals are speculative and probably inordinately expensive.

The "hydrogen economy" dream is a mixture of two different goals: to use hydrogen to store renewable energy; and more immediately, to find a replacement for gasoline. The idea is to extract hydrogen from water and use it as a fuel. Fed into fuel cells (when they become practical), it would provide clean power more efficiently than the internal combustion engine. And, in this dream, the electricity could be fed into the power grid, thus solving the problem of intermittency.

That seems reasonable, at first. Hydrogen is the most common element in the known universe. The Secretary of Energy once exulted that "hydrogen is free!" But that is far from true. We are learning some of the problems. Water (H_2O) is a tenacious molecule, and nobody has yet found a way to split it and sequester the hydrogen except with energy inputs far exceeding the potential output. That means that hydrogen is not a *source* of energy unless someday we should learn to extract it from water with a net energy gain. Failing that, it is simply an energy *carrier*—a way to store energy until it is needed. And it is an intractable element. It is the lightest of gases, and a huge volume holds little energy compared to conventional energy sources. To use it, particularly for transportation, it must be liquified by cooling it to nearly absolute zero, and keeping it there—which may take 30 percent of the energy in the hydrogen—or stored under pressures up to 10,000 psi (pounds per square inch), which also requires a lot of energy and demands an enormously strong tank much larger than present gasoline tanks, which converts every vehicle into a mobile bomb. Hydrogen is difficult to transport and prone to leakage, which of course dissipates the energy gained and which may eventually become a problem in the atmosphere. It would be hard to transport, since it turns metals brittle. It ignites at a dangerously low temperature.

We can indeed isolate hydrogen. It is already produced for specialized industrial uses. But it is expensive. It wholesales in my state for $98 per 1000 cubic feet, in cylinders. By contrast, the utility company delivers that much natural gas for $6. The price ratio is 16:1, which is over 50:1 in energy terms. Only a small part of that differential results from different scales and markets. And that hydrogen is steam stripped from natural gas. Generating hydrogen by electrolysis—which will be the process when natural gas is gone—is three times as expensive.

Most experts seem to have become disenchanted with the hydrogen dream. Two recent reports almost ignore the goal of energy storage for the grid because of the energy losses in the multiple conversions from wind or solar energy to electricity to hydrogen to electricity. They concentrate on the more immediate task of using hydrogen as a replacement for gasoline, and even there they caution against too much optimism. A National Academy of Engineering (NAE) report concludes that hydrogen can be extracted from coal and used to power vehicles, at something over twice the present price of gasoline, but concludes that the use of renewables, "except possibly wind" for hydrogen production at competitive prices, awaits new technologies.[12] Staff writer Robert F. Service in Science recently remarked that, "A hydrogen economy, if it comes at all, won't happen soon," and Ernest Moniz (an MIT physicist and former Undersecretary of Energy) agreed that, "It's very, very far away . . . Let's just say decades, and I don't mean one or two."[13] Other writers are, if anything, more skeptical. Andrew Ferguson concludes that neither wind nor biomass will be viable sources of hydrogen.

Those experts are all addressing the effort to preserve the present energy supply, at prices not too far above present levels. That probably is not an option in coming decades, and it won't be an option in the post-fossil era. Moreover, if a way is not found to store energy inexpensively, wind and solar power will never provide a reliable energy supply. Future societies may have to adapt to accommodate periodic and unpredictable blackouts and sustained shortages in long cloudy and windless periods. It may not be our world, but the world would survive.

Scientists are working at the task of making hydrogen an economic energy carrier. They are trying everything from making electrolysis less costly by tuning the wavelength of energy, to harvesting pond scum (which releases minute quantities of atomic hydrogen), to storing the hydrogen in metal hydrides. They are a long way from the goal.

THE LONG SHOTS

In theory, one can generate electricity from tidal energy, ocean waves and currents, geothermal vents, and even the temperature gradient in the deep tropical oceans. Most of them are intermittent, but some are predictable. **Geothermal** energy has been a disappointment. Iceland, with its volcanic activity, gets its electricity from geothermal energy, but at the biggest site in the United States (the Geysers field in California), production has been declining 10 percent per year because of the loss of groundwater to produce steam for the boilers. An international experiment in New Mexico in the 1970s ("dry hot rocks") undertook to pump in water to generate steam, but the experiment fizzled because the water dissipated.

Experiments with **ocean wave energy** have been wrecked by storms or produced only enough energy for small local projects. Opinions differ remarkably as to its potential, but it will be a costly and dispersed source of energy. In an **ocean current** experiment in Norway, a turbine has been placed in a strait with strong currents; it may fare better, but there are not many such locations to exploit.

The WEC lists twenty-seven sites, worldwide, that have been studied for **tidal power** It is a capital-intensive energy source. Most of the high-potential sites are in Russia, where three sites can perhaps support a total of 110,000 megawatts of generating capacity. Since the capacity factor is about 25 percent, the actual production would be more like 27,000 megawatts, but that is the equivalent of about thirty fossil fuel plants. However, that list probably exhausts the major potential sites.

THE LONGER SHOTS

Fusion power is seen by its advocates as the deus ex machina. It would run on deuterium (heavy hydrogen, which is abundant in seawater) and avoid most of the pitfalls of fission. The problem is that it is very difficult indeed to maintain a continuing process of fusion at about 100 million degrees Fahrenheit, contained only by an extraordinarily intense magnetic field. After nearly fifty years of efforts, scientists have succeeded in containing the fusion on a laboratory scale, and they have even claimed success in briefly generating nearly half as much energy as they were putting into the magnetic containment. The challenge is to show that they can generate net energy, safely and on a sustained basis, scale it up, and harness it to generate electricity. The next step in the search is a proposed international experiment called the ITER. As of now, the putative sponsors have yet to resolve a bitter contest as to whether the experiment will be located in France or Japan. The path toward fusion energy has not been smooth, and there is no way of knowing whether it will ever work.

If fusion becomes workable and inexpensive, it will rewrite the whole transition scenario. With almost limitless electricity available, at no cost to the environment or the climate, we will reenter an era of plenty, unconstrained by the limited annual input of sunlight and with none of fossil fuels' penalties.

Let me inject a brief aphorism: Solutions beget problems. The human race would face an entirely new set of issues. How will our economies be transformed when almost all energy is electricity, which provides neither feedstocks nor powerful, highly portable propulsion? Will the gap between the haves and have-nots intensify because poor nations and people will not be partners in this extraordinarily complex technology? Will the human race in its hubris be able to restrain its enthusiasm for cheap power? If it does not, will we drive our numbers and consumption to levels that imperil us because of the other penalties of growth? Will we imperil the other creatures on a shared planet—and eventually our own survival?

Clathrates For completeness, let me mention the last faint hope of those who cling to the fossil era. Clathrates are globules of ice and methane, widely distributed on the continental shelves of the world's oceans, and in the tundra. It is a highly dispersed resource. To disturb the clathrates is to pose the danger of releasing the methane (which is a potent greenhouse gas) or possibly generating mudslides and tsunamis on the sloping continental shelf in the mining process, and large scale dredging would inflict damage on marine fisheries beyond anything that fishing trawlers have yet inflicted.

The scientific consensus seems to be that clathrates are a highly uncertain and potentially disruptive source of energy; but Japan, in its tireless search for energy, has led at least two expeditions to see if the methane can be captured.

THE ENERGY MIX OF THE FUTURE

Prof. Ted Trainer (Note 4), after an exhaustive examination of solar, wind, and biomass energy, and hydrogen storage, concluded that:

> " . . . implausible assumptions would have to be made before it could be concluded that present electrical and liquid fuel

demand could be met from solar sources, let alone demand anticipated in view of continued economic growth. In other words renewable energy forma are unlikely to be capable of substituting for fossil fuels, and the shortfall is likely to be large, especially with respect to liquid fuels.

"It should be emphasized again that the foregoing argument does not imply that renewable energy sources should be rejected We should change to them as rapidly as possible, and we could live well on them but only if we accept transition to a very different society."

That's a pretty good statement of the issue. He might have added that decreasing population and lower demand would make the problem easier.

The broad outlines of the post-fossil energy mix may be clearer than the details.

Biomass is the only candidate that can duplicate most of fossil fuels' roles, but its availability will be determined by future population size. A halving of world population would make it possible for biomass to provide much less than a quarter of present world energy use—but then again, energy needs would be much less than they are now. A world population of 1.6 billion— the 1900 level—would have more food per capita, and the poor majority would have access to more energy than they have now.

Wind can augment that energy, but good sites are not unlimited, and wind will not fill the fossil fuel gap. Direct solar can play a multiple role—perhaps more focused on distributed local energy and the better use of sunlight than grid energy. Direct solar energy does not have the inherent limits of the other sources, but its availability will be a function of its price and the land it requires.

Multiple local sources such as wave energy and ocean currents can further augment the supply, but sites that can pay back the energy invested in them are not common.

Most of the energy will be electricity, and that will change our economies. So will the sporadic nature of much of the energy, unless a satisfactory storage system is found, and that will put a premium on flexible, independent, and small-scale energy systems.

All of the renewables, even by the most optimistic predictions, will be dramatically more expensive than today's energy. And—unless there is fusion energy, which is still a dream—the available energy will be far less than at present. It would be an insuperable task to try to replace current sources at current scales.

The debate has been cast in the wrong terms. The problem cannot be solved if we keep asking: "What energy sources will be available to replace fossil fuels?" We should instead ask: What populations can be supported at a decent standard by the energy sources that will be available after the transition from fossil fuels? There are such sources, but they won't be like the profligate fossil fuel era.

Talking about lifestyles is always tricky terrain. It can lead to maundering generalizations about the good old days, but I am serious in suggesting that we will need to adjust our goals to seek a simpler and less energy-dependent pattern of living. The models are there from the human experience, and the new technologies can improve upon them.

THE PROBABLE IMPACTS OF THE TRANSITION

Agriculture. Modern agriculture has been called a process of using sunlight to convert hydrocarbons into food—very inefficiently. In the United States, we use about ten calories of fossil fuels, in the form of fertilizers, pesticides, and powered machinery, to produce one calorie of food. (That understates the total, since it does not include the off-farm energy.) However, we are remarkably efficient in terms of labor. Giampietro and Pimentel have calculated that it takes 166 times as much labor to raise grain in China as in the United States.[14] Even with all that fossil energy, U.S. agriculture is encountering problems: declining arable land and water, faltering yields, resistant pests. American agriculture's reliance on fossil energy during the energy transition is a catastrophic mistake. We need to find ways, as fossil fuels decline, to work the land efficiently with less energy and limited labor during a transition to lower population levels. The alternative, importing more labor from overpopulated countries, would vitiate any effort to reduce U.S. population—even as migration of people with high fertility is driving U.S. population upwards now. It is a dilemma, but we can profit from the opportunity that comes with a smaller population, to bring our national water budget in balance and to concentrate our effort on the better land.

We have separated animal husbandry from crop production. Grain production depends now on commercial fertilizers. Meanwhile, more and more chickens and hogs and cattle are raised in huge "agrifactories," ten thousand or more at a time. Their manure is held in large holding ponds, from whence it escapes and flows toward the sea. The "dead zones" in the Gulf of Mexico, the Chesapeake Bay, and off the Massachusetts and California coasts are the product of excess chemical fertilizers and manure running into the sea, and the resultant eutrophication is destroying fisheries. Hurricane Floyd in 1999 overwhelmed the hog farm holding ponds in North Carolina and washed the manure into the Albemarle and Pamlico Sounds, with impacts on fisheries and marine life that are still being felt.

The system is fundamentally out of whack. The animals should be raised where the natural fertilizers they produce can be used on the crops. It would eliminate two sources of pollution, save energy, and save the fisheries. I have run a manure spreader. Hardly glamorous work, but necessary. You don't see them anymore, but they will be back. Diminishing fossil energy will drive up the price of chemical fertilizers. When the fossil fuels are gone, it will make no sense to harvest biomass to make commercial fertilizer when better fertilizer is already available on the farm. Some of this reversion to old practices will be driven by prices, but it would move much faster if agricultural producers were charged the pollution costs from excess commercial fertilizer and manure holding ponds.

The Chinese carry it a step further and use human waste, nightsoil, to produce biogas and fertilizer. Chinese cities are ringed by highly productive truck farms. Fermentation can kill the bacteria and make the process safe—but China is terribly crowded. In America, dried sewage sludge is sometimes used for fertilizer, but not generally on food crops, and we do not use the biogas.

U.S. corn yields in the early 1900s were about one-quarter of present yields; wheat yields were one-third. With modern agronomy, we could do better than that. We could reserve commercial fertilizer inputs for nutrients, particularly phosphorus, that tend to go down with repeated harvesting. It is difficult to put a number on the appropriate U.S. population size for such a farm economy, but we supported 100–125 million people on that economy. We could probably support 125–150 million people, better, and still have enough land to produce the biomass we need. And it would be a system in balance.

Europe, Japan, and the less developed countries will be making their own calculations, and they are much grimmer.

Transportation and Communication. The energy transition will profoundly change transportation and the way we live. There are now more private vehicles than adults in the United States; energy costs will drive us back to public transport. Airplanes will begin to disappear because no fuel substitute is as cheap and energetic as petroleum. Electricity and (briefly) coal will give trains an advantage over trucks. Shipment by sea, powered by coal, nuclear—or wind—will regain the competitive advantage it had 150 years ago. Half of ocean shipping tonnage now is used to ship petroleum, and that will stop. The limits on renewable energy will force a return to more local patterns of living, working, and shopping. Locally grown food will enjoy a widening price advantage over distant supplies. Telecommunications are energy-efficient; they will make increasing inroads into business and personal travel.

Industry. Higher energy costs and shifting sources will drive wholesale adjustments in the economy. For example, smelted metals and building materials will get much more expensive, driving up the demand for lumber. Pharmaceuticals, plastics, and textiles will go back to biomass and cellulose feedstocks. The cost of retooling will be considerable. Since factories' life cycle is about thirty years, those industrialists who see the future will plan the transition early and gain an inestimable advantage over those who don't.

Governments have a role in fostering the scientific inquiry that will help those industrialists identify the right choices.

Consumption and Equity. An effort by the rich countries to maintain present production patterns, and even more the vision of growth as a solution, will be suicidal. We will need instead a policy of accommodating the changes. The larger the population that has to be accommodated, the greater the investment needed.

The United States is regularly criticized for its high consumption levels. That particular criticism will take care of itself. High-energy prices will lead to lower material standards of living, though perhaps not of real well-being.

Will the readjustment be spread evenly? A shrinking pie intensifies the competition. Income differentials that were tolerable may become the stuff of class warfare.

Toward a More Benign System. Renewables will end the present insults to the environment, and they can be a boon if a new population transition lowers the pressures on resources. They will not affect the climate. (Even biomass absorbs and emits carbon in short-term cycles.) That will not be the end of climate warming because past and foreseeable carbon emissions will drive the warming trend for centuries. But at least it will bring the end of that warming in sight. In the title, I referred to "dawn." Perhaps we can see the end of a potentially suicidal trajectory and the beginning of a time in which humans can live sustainably on Earth.

THE SOLUTION ON THE DEMAND SIDE

One of cartoonists' stock figures is a little man with a beard, in a robe and sandals, standing on a street corner holding up a placard with the message, "Repent, for the end is near!" From time to time, I think of that little guy and ask myself, is that me? On reflection, however, I assure myself that my concerns are real. Even prophets are occasionally right. The scientists who are worried are the ones who present the weight of evidence, and their opponents generally resort to the mantras of faith, such as "They'll find more resources," or "The market will adjust." (which is hardly a solution) or "Science will find a solution." I don't believe them.

The Two Child Family. Population growth set the scene for the overshoot scenario, and a reversal of that growth will be necessary to get us out of it. We cannot go back to the renewable energy economy of 1900 without a much smaller population. That reversal will be voluntary and manageable if we are wise, and catastrophic if we are not. It is not fated to be painful. We think of hungry peasants doing subsistence farming, but their poverty is a function of too many people on too little land. Think instead of the Amish, who work hard but live well in a largely self-contained, renewable energy system—but they have good land, and enough of it. (They also have large families, which raises the question: Will fertility rise as urbanized societies de-urbanize, and children again become an economic asset rather than a burden? But I'll save that exploration for a later time.)

In 1994, I wrote an NPG FORUM called "The Two Child Family." The point was that stopping with two children would be a painless way to stop and reverse population growth. Because some women have no children, or one, a two child maximum would mean a total fertility rate of about 1.5. It would not be an onerous limit. More than 70 percent of American women stop by then, even now. Let me update the graph in that article.

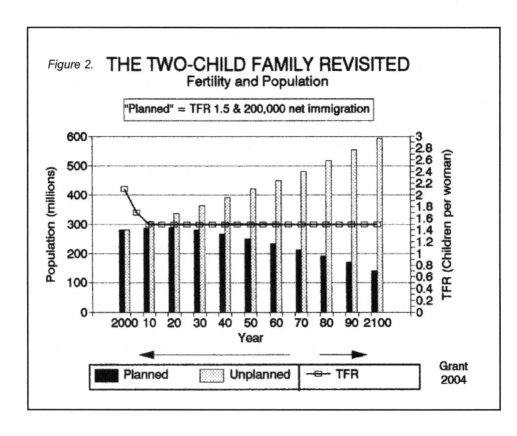

Figure 2. **THE TWO-CHILD FAMILY REVISITED**
Fertility and Population

"Planned" = TFR 1.5 & 200,000 net immigration

I had to revise the original chart. The numbers got out of date. The unplanned figure for 2100 (based on the Census Bureau middle projection) is now 100 million more than the 1994 projection. That is a sobering change; it shows how much faster we are growing than we thought, then. We underestimated the impact of mass immigration. The need to bring population down has become more urgent and more difficult, so unlike the first graph, this one does not show population stabilization during this century. It must occur, and not too long after that, to stabilize at a better level.

The graph is an idealization. It, too, will be overtaken if women do not decide to "stop at two." We need a consensus that population must come down, plus the political and social will to act. I wonder whether our modern, individualistic society can act in so cohesive a fashion.

That new demographic transition is underway in Europe, Japan, South Korea, and Taiwan, where the women (if not the governments) have shown the way to smaller populations through fewer children, but they seem not to have learned yet that too few leads to extinction. Many of the so-called "less developed countries" are trying to stop population growth, but they still seek industrialization on the American pattern. That leads to a dead end, as we are just beginning to realize. They would be well advised to reconsider that objective.

The fossil fuel bubble was a durable one, and unlike soap bubbles, it will collapse slowly. That gives the world some time to make the one real accommodation that will provide a smooth transition to the leaner times ahead: a deliberate policy of negative population growth.

I do not mean to understate the difficulties, nation by nation, of learning to manage population size to maximize human welfare, or the potential for conflict as different nations move at different rates toward sustainable population levels, or the tensions created as crowded nations eye other nations' land, water, or resources. I would not bet that the human race can manage this most difficult of transitions—this retreat from overshoot—without turmoil. But we have an opportunity to try.

NOTES

Chapter 1

1. www.newamericancentury.com

2. "National Security Strategy of the U.S.A.," transmitted by President Bush to Congress on 9-18-2002.

3. Lewis E. Lehrman, Co-Chairman of PNAC, "Energetic America," in the *Weekly Standard*, September 23, 2003. See PNAC Web site Note 1.

4. Environmental News Service, 4-8-2003; full text at www.state.gov.

5. David Herlihy, *The Black Death and the Transformation of the West* (Cambridge, MA: Harvard University Press, 1997).

6. Data from U.S. Department of Agriculture, *World Agriculture*, Statistical Bulletin 861, November 1993. In the "new world," I include the Western hemisphere, Australia and New Zealand. "Europe" excludes Russia.

7. "A great transition in human history will have begun when civilized man endeavors to assume conscious control (of population growth) in his own hands, away from the blind instinct of mere predominant survival." J.M. Keynes, Preface to Harold Wright, *Population* (London: Harcourt Brace, 1923).

8. Keynes' famous statement that "Avarice and usury and precaution must be our gods a little longer still." is quoted in E.F. Schumacher, *Small is Beautiful: Economics as if People Mattered* (London: Blond and Briggs, 1973, reprints by Harper & Row, New York, 1975 to 1989; p.24).

9. George Perkins Marsh, *Man and Nature, Or, Physical Geography as Modified by Human Action* (originally published 1864. Reprinted Cambridge: Harvard University Press, 1965).

10. Joint Statement by the Presidents of the U.S. National Academy of Sciences and the British Royal Society, released February 26, 1992, by the National Research Council, Washington.

11. Paul D. Gottlieb, "Growth Without Growth: An Alternative Economic Development Goal for Metropolitan Areas" (Washington: Brookings Institution Discussion Paper, February 2002, p. 25).

12. Most famous is E.F. Schumacher, *Small is Beautiful* (op cit). The evidence that the working classes have enjoyed little or, in some periods, none of the benefits of growth has been assembled by Richard Douthwaite in *The Growth Illusion* (Tulsa, OK: Council Oak Books, 1993; originally published in Great Britain by Green Books, 1992).

13. For citations and a much fuller exploration of these limits, see Lindsey Grant, et al, *Elephants in the Volkswagen* (New York: W.H. Freeman, 1992), Lindsey Grant, *Juggernaut: Growth on a Finite Planet* (Santa Ana: Seven Locks Press, 1996); *Too Many People: The Case for Reversing Growth* (Seven Locks Press, 2000); and "Diverging Demography, Converging Destinies" (Alexandria, VA: Negative Population Growth, Inc. FORUM series January 2003; also at www.npg.org).

14. The U.S. Geological Survey estimates remaining world resources at 2269 billion barrels. (USGS, Digital Data Series DDS-60) It is one of the more optimistic projections, and will be discussed at length in Chapter 2.

15. The EIA, in its *International Energy Outlook 2003*, projects world oil demand to rise by 41 million barrels per day—53 percent—by 2025, and 15 million barrels of that growth will be in developing Asia.

16. *U.S. Statistical Abstract 2001*, Table 877.

17. U.S. Department of Energy, Energy Information Administration (EIA), "Monthly Energy Review, March 2004," Table 11.2.

18. Walter L. Youngquist, GeoDestinies (Eugene, OR: National Book Company, 1997, Chapter 13). Richard A. Kerr, "Gas Hydrate Resource: Smaller But Sooner," *Science*, Vol. 303, 2-14-04, pp.246–247.

19. With the lower populations of 1950, total U.S. emissions would be 54 percent of the present 1.57 billion metric tons. Emissions by the rest of the industrial world would be 73 percent of 2.34 billion tons. Developing country emissions would be 35 percent of 2.53 billion tons. Totaled, world emissions from fossil energy would be 53 percent of the present 6.44 billion tons. (Data from U.N. Population Division and U.S. Department of Energy, Energy Information Agency, *International Energy Annual, 2000.*) Moreover, lower populations would mean less destruction of tropical forests, which presently add roughly 20 percent to world greenhouse gas emissions.

20. I treat this phenomenon at length in "It's the Numbers, Stupid," NPG FORUM September 2003. See www.npg.org.

21. The discussion of optimum population in this paper expands on an article titled "Optimum Population: How Many Is Too Many?" published in the August-September 2004 issue of *FREE INQUIRY, the Journal of the Council for Secular Humanism.*

Chapter 2

1. 2000 Census Bureau middle projection 2000–2100, modified by 2002 interim projection 2000–2050 and my adjustments 2060–2100.
2. Bureau of the Census, *Historical Statistics of the United States: Colonial Times to 1957*, Table M71-87, and *Statistical Abstract of the United States, 2003,* Table 895; U.S. Department of Energy, Energy Information Administration (DOE/EIA), *Annual Energy Report 2002*, Table 1.3.
3. DOE/EIA *Annual Energy Report 2003*, Table 2.1a "Energy Consumption by Sector, 1949–2002."
4. DOE/EIA unnumbered table "Feedstock Energy . . . " dated 1998, and ww.eia.gov/emeu/mecs/trends/feedstock.htm.
5. Executive Summary, USGS World Energy Assessment 2000, Digital Data Series 60, Table 1. The Assessment gives 5 percent, 50 percent, mean, and 95 percent confidence levels for undiscovered oil resources and reserve growth. The mean estimates for undiscovered resources are slightly higher than the 50 percent confidence levels: 649bb for oil, 4669tcf for gas, and 207bb for natural gas liquids. Its figures for natural gas liquids are included in the U.S. petroleum estimates but not in the rest of the world figures. I have included them in petroleum throughout.
6. Jeff Garth & Stephen Labaton, "Oman's Oil Yield Long in Decline, Shell Data Show," *New York Times*, 4/8/04.
7. DOE/EIA, *International Energy Outlook 2004*, Table A2. "World Total Energy Consumption by Region & Fuel, Reference Case." EIA is an excellent source for past and current data but carries its projections only to 2025, and they show no evidence that the possibility of resource constraints was considered.
8. Colin J. Campbell, "Forecasting Global Oil Supply," *Submission to H.M. Government Consultation on Energy Policy*, undated,(2003); A.M.S. Bakhtiari (National Iranian Oil Company), "World Oil Production Capacity Model Suggests Output Peak by 2006–07," *Oil & Gas Journal*, April 26, 2004, pp.18–20; Richard C. Duncan & Walter Youngquist, "Encircling the Peak of

World Oil Production," in International Association for Mathematical Geology *Natural Resources Research*, Vol.8, No.3, 1999.

9. From Colin J. Campbell's projection of a plateau 2015-2040 followed by a sharp and unheralded decline, in "Submission to H.M. Government Consultations on Energy Policy," (above), pp.6–8. www.dti.gov.uk/energy/energyep/index.com.

10. DOE/EIA International Energy Outlook 2004.

11. Walter Youngquist, "Spending Our Great Inheritance—Then What?" *Geotimes*, July 1998.

12. Walter Youngquist, "Shale Oil—The Elusive Energy," in *Hubbert Center Newsletter 98-4*, Colorado School of Mines, Golden, Colorado.

13. *Statistical Abstract of the United States 2003*, Table 891; International Energy Outlook 2004, www.eia.doe.gov/oiaf/ieo/coal.html.

14. World Energy Council (WEC), *2001 Survey of Energy Resources* This is the most comprehensive synoptic survey of world energy. The figures it gives for unproven coal resources are so incomplete and the assumptions so various that they are best ignored in this article. *The BP Statistical Summary of World Energy 2004* coal projections are drawn from the WEC figures.

15. Statistical Abstract, op cit, Table 1363.

16. Personal communication July 8, 2004 from Neville Holt, Electric Power Research Institute, Palo Alto, CA.

17. Paul Roberts, *The End of Oil* (Boston: Houghton, Mifflin, 2004), p.269.

18. Paul B. Weisz, "Basic Choices and Constraints on Long Term Energy Supplies," Vol. 57, 7/04, www.physicstoday.org/vol-57/iss-7/p47.html.

19. Thomas Hirons, Los Alamos National Laboratory, personal communication July 2, 2004.

20. Paul G. Falkowski et al, "The Evolution of Modern Eukaryotic Phytoplankton," *Science*, Vol. 305, 7-16-2004, p.354ff, Fig.3.

Chapter 3

1. See World Energy Council (WEC), *2001 Survey of World Energy Resources, Wood (including Charcoal)*, text and Table 9.1. It cites an International Energy Agency (IEA) estimate for all biomass, which implies a figure of 9 percent.

2. U.S. Bureau of the Census, *Statistical Abstract of the United States, 2003,* Table 899.

3. WEC, see note 1.

4. Trainer, Ted, Faculty of Arts, Univ. of N.S.W., Australia, "Renewable Energy: What Are the Limits?" circulated by email. See www.arts.unsw.edu.au/tsw/ D74.RENEWABLE-ENERGY.html. Trainer's study summarizes the conclusions of David Pimentel and others on net energy yields from grains and makes clear the wide variety of opinions among the experts.

5. The most cited estimate is that of Vitousek, P.M. et al., "Human Appropriation of the Products of Photosynthesis," *Bioscience*, June 1986, pp.750–760: that humans appropriate about 40 percent of terrestrial net photosynthetic energy.

6. Cited by Trainer (Note 4).

7. An American Wind Energy Association (AWEA) estimate quoted by Andrew Ferguson, Research Director, Optimum Population Trust, UK (www.optimumpopulation.org), suggested that the limit would be about 22 percent of present U.S. electricity consumption. ("Verdict on the Hydrogen Experiment: An Update," draft article dated 10-20-02; personal communication.) I am indebted to Mr. Ferguson for much advice on the costs of renewable energy.

8. The ratio seems high, given other prices cited; it may include an estimate of how much it would cost to store the energy, which is still an unknown.

9. Email 6-29-04 from Andrew F. Ferguson (Note 7).

10. Howard C. Hayden, The Solar Fraud (Pueblo, CO: Vales Lake Publishing, 2001), p.161.

11. Hayden (Note 10), p.154.

12. National Academy of Engineering, Board on Energy & Environmental Systems, *The Hydrogen Economy: Opportunities, Costs, Barriers and R&D Needs* (Washington: National Academies Press, 2004), Executive Summary p.2.

13. Science, 8-13-04, p. 958. The series of articles misleadingly entitled "Toward a Hydrogen Economy," pp.957–976, is a useful summary of the difficulties of using hydrogen for energy.

14. Mario Giampietro & David Pimentel, "The Tightening Conflict: Population, Energy Use and the Ecology of Agriculture," NPG FORUM series, Oct. 1993.

About the Author

Lindsey Grant writes on population and public policy. A retired Foreign Service Officer, he was a China specialist and served as Director of the Office of Asian Communist Affairs, National Security Council staff member, and Department of State Policy Planning Staff member.

As Deputy Assistant Secretary of State for Environment and Population Affairs, he was Department of State coordinator for the Global 2000 Report to the President, Chairman of the interagency committee on International Environmental Affairs, United States delegate to (and Vice Chairman of) the OECD Environmental Committee and United States member of the UN ECE Committee of Experts on the Environment.

His books include: *Too Many People: The Case for Reversing Growth, Juggernaut: Growth on a Finite Planet, Foresight and National Decisions: the Horseman and the Bureaucrat, Elephants in the Volkswagen* (a study of optimum United States population) and *How Many Americans?* He edited the forthcoming book, *The Case for Fewer People: The NPG Forum Papers.*

About NPG

Negative Population Growth, Inc. was founded in 1972. In the same year, thirty-four of Great Britain's most distinguished scientists, including Sir Julian Huxley, endorsed the basic principles of a landmark study called *A Blueprint for Survival.* The study warned that demand for natural resources was becoming so great that it would exhaust reserves and inevitably cause the breakdown of society and the irreversible destruction of the life-support systems on this planet. To prevent disaster they urged Britain to cut its population in half. That study greatly influenced the thinking of NPG's founders, who were convinced that a similar reduction in population was necessary for our own country.

Now, thirty-two years later, NPG is a fast-growing organization with over 30,000 members nationwide. We continue our efforts to better educate the American people and our policymakers about the devastating effects of over-population on our resources, environment and the quality of our lives. Our purpose, broadly stated, is, through public education, to encourage the United States, and then every country in the entire world, to put into effect national programs with the goals of first achieving a negative rate of population growth, then eventually stabilizing population size at a far lower level than today's—a size that would be sustainable indefinitely in a sound and healthy environment.

Our extensive publications are intended to both stimulate public discussion and debate, and to change the way our opinion leaders and policymakers view the issues of population size and growth, and their impact on our environment, resources and quality of life.